# Geotechnical Engineering and Soil Science

# Geotechnical Engineering and Soil Science

Edited by
**Greg Powell**

Larsen & Keller
www.larsen-keller.com

Geotechnical Engineering and Soil Science
Edited by Greg Powell
ISBN: 978-1-63549-644-4 (Hardback)

≡ Larsen & Keller

Published by Larsen and Keller Education,
5 Penn Plaza,
19th Floor,
New York, NY 10001, USA

**Cataloging-in-Publication Data**

Geotechnical engineering and soil science / edited by Greg Powell.
    p. cm.
Includes bibliographical references and index.
ISBN 978-1-63549-644-4
1. Geotechnical engineering. 2. Soil science. 3. Engineering geology. 4. Soil mechanics.
I. Powell, Greg.
TA703.5 .G46 2018
624.151--dc23

For more information regarding Larsen and Keller Education and its products, please visit the publisher's website www.larsen-keller.com

# Table of Contents

# Preface

Geotechnical engineering uses the knowledge of soil science to understand the behaviour of earth materials. As soil science, is the study of the nature of soil, along with its chemical, physical, fertility, and biological properties, it plays a crucial role in geotechnical engineering in understanding soil mechanics with respect to construction. Geotechnical engineers use this knowledge for civil engineering projects. This book is designed to provide in-depth information about this subject. The topics included in it on geotechnical engineering and soil science are of utmost significance and bound to provide incredible insights to readers. This textbook will serve as a valuable source of reference for those interested in this field.

A detailed account of the significant topics covered in this book is provided below:

Chapter 1- The engineering behavior of Earth materials is known as geotechnical engineering. It uses principles of soil mechanics and rock mechanics to study the soil. Risk assess-ments are also done. The chapter on geotechnical engineering and soil science offers an insightful focus, keeping in mind the complex subject matter.

Chapter 2- Clay is formed with the help of water and is a major element in the speculative theories regarding the origin of life. They are phyllosilicates and are found as sheet structures. Clay is found on Earth as well as other planets. The major categories of clay are dealt with great details in the chapter.

Chapter 3- Soils contain many interconnected pores to provide a route for water to pass. The mean value of the rate of flow is referred to as permeability. This is best understood by Darcy's Law, which provides an equation to calculate rate of flow through pores. The chapter also explores other related concepts such as Navier–Stokes equations, measurement of permeability, etc. The topics discussed in the chapter are of great importance to broad-en the existing knowledge on geotechnical engineering and soil science.

Chapter 4- Shear strength is the stress felt on the cross-section of a surface. Shear strength of soil analyzes the shear stress that soil can maintain. This chapter serves as a source to un-derstand the major categories related to soil texture. Geotechnical engineering is best understood in confluence with the major topics listed in the following chapter.

Chapter 5- The process by which the volume of soil decreases is known as is called soil consolida-tion. An oedometer test can analyze the compression of soil. One-dimensional consoli-dation and secondary consolidation have also been mentioned.

The aspects elucidated in this chapter are of vital importance, and provide a better understanding of geotech-nical engineering and soil science.

It gives me an immense pleasure to thank our entire team for their efforts. Finally in the end, I would like to thank my family and colleagues who have been a great source of inspiration and support.

**Editor**

# An Introduction to Geotechnical Engineering and Soil Science

The engineering behavior of Earth materials is known as geotechnical engineering. It uses principles of soil mechanics and rock mechanics to study the soil. Risk assessments are also done. The chapter on geotechnical engineering and soil science offers an insightful focus, keeping in mind the complex subject matter.

## Geotechnical Engineering

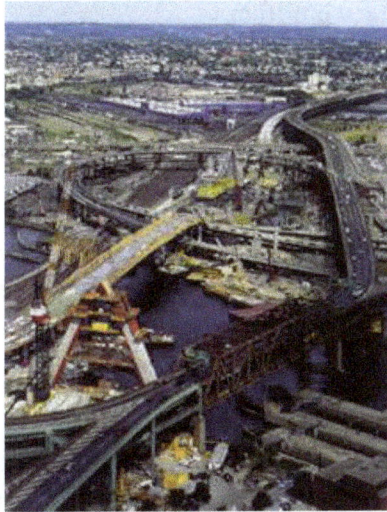

Boston's Big Dig presented geotechnical challenges in an urban environment.

Geotechnical engineering is the branch of civil engineeringconcerned with the engineering behavior of earth materials. Geotechnical engineering is important in civil engineering, but also has applications in military, mining, petroleum and other engineering disciplines that are concerned with construction occurring on the surface or within the ground. Geotechnical engineering uses principles of soil mechanics and rock mechanics to investigate subsurface conditions and materials; determine the relevant physical/ mechanical and chemical properties of these materials; evaluate stability of natural slopes and man-made soil deposits; assess risks posed by site conditions; design earthworks and structure foundations; and monitor site conditions, earthwork and foundation construction.

A typical geotechnical engineering project begins with a review of project needs to define the required material properties. Then follows a site investigation of soil, rock, fault distribution and bedrock properties on and below an area of interest to determine their engineering properties including how they will interact with, on or in a proposed construction. Site investigations are needed to gain an understanding of the area in or on which the engineering will take place. Investigations can include the assessment of the risk to humans, property and the environment from natural hazards such as earthquakes, landslides, sinkholes, soil liquefaction, debris flows and rockfalls.

A geotechnical engineer then determines and designs the type of foundations, earthworks, and/or pavement subgrades required for the intended man-made structures to be built. Foundations are designed and constructed for structures of various sizes such as high-rise buildings, bridges, medium to large commercial buildings, and smaller structures where the soil conditions do not allow code-based design.

Foundations built for above-ground structures include shallow and deep foundations. Retaining structures include earth-filled dams and retaining walls. Earthworks include embankments, tunnels, dikes and levees, channels, reservoirs, deposition of hazardous waste and sanitary landfills.

Geotechnical engineering is also related to coastal and ocean engineering. Coastal engineering can involve the design and construction of wharves, marinas, and jetties. Ocean engineering can involve foundation and anchor systems for offshore structures such as oil platforms.

The fields of geotechnical engineering and engineering geology are closely related, and have large areas of overlap. However, the field of geotechnical engineering is a specialty of engineering, where the field of engineering geology is a specialty of geology.

## History

Humans have historically used soil as a material for flood control, irrigation purposes, burial sites, building foundations, and as construction material for buildings. First activities were linked to irrigation and flood control, as demonstrated by traces of dykes, dams, and canals dating back to at least 2000 BCE that were found in ancient Egypt, ancient Mesopotamia and the Fertile Crescent, as well as around the early settlements of Mohenjo Daro and Harappa in the Indus valley. As the cities expanded, structures were erected supported by formalized foundations; Ancient Greeks notably constructed pad footings and strip-and-raft foundations. Until the 18th century, however, no theoretical basis for soil design had been developed and the discipline was more of an art than a science, relying on past experience.

Several foundation-related engineering problems, such as the Leaning Tower of Pisa, prompted scientists to begin taking a more scientific-based approach to examining the subsurface. The earliest advances occurred in the development of earth pressure theories for the construction of retaining walls. Henri Gautier, a French Royal Engineer, recognized the "natural slope" of different soils in 1717, an idea later known as the soil's angle of repose. A rudimentary soil classification system was also developed based on a material's unit weight, which is no longer considered a good indication of soil type.

The application of the principles of mechanics to soils was documented as early as 1773 when Charles Coulomb (a physicist, engineer, and army Captain) developed improved methods to determine the earth pressures against military ramparts. Coulomb observed that, at failure, a distinct slip plane would form behind a sliding retaining wall and he suggested that the maximum shear stress on the slip plane, for design purposes, was the sum of the soil cohesion, $c$, and friction $\sigma \ \tan(\phi)$, where $\sigma$ is the normal stress on the slip plane and $\phi$ is the friction angle of the soil. By combining Coulomb's theory with Christian Otto Mohr's 2D stress state, the theory became known as Mohr-Coulomb theory. Although it is now recognized that precise determination of cohesion is impossible because $c$ is not a fundamental soil property, the Mohr-Coulomb theory is still used in practice today.

In the 19th century Henry Darcy developed what is now known as Darcy›s Law describing the flow of fluids in porous media. Joseph Boussinesq (a mathematician and physicist) developed theories of stress distribution in elastic solids that proved useful for estimating stresses at depth in the ground; William Rankine, an engineer and physicist, developed an alternative to Coulomb's earth pressure theory. Albert Atterbergdeveloped the clay consistency indices that are still used today for soil classification. Osborne Reynolds recognized in 1885 that shearing causes volumetric dilation of dense and contraction of loose granular materials.

Modern geotechnical engineering is said to have begun in 1925 with the publication of *Erdbaumechanik* by Karl Terzaghi (a civil engineer and geologist). Considered by many to be the father of modern soil mechanics and geotechnical engineering, Terzaghi developed the principle of effective stress, and demonstrated that the shear strength of soil is controlled by effective stress. Terzaghi also developed the framework for theories of bearing capacity of foundations, and the theory for prediction of the rate of settlement of clay layers due to consolidation. In his 1948 book, Donald Taylor recognized that interlocking and dilation of densely packed particles contributed to the peak strength of a soil. The interrelationships between volume change behavior (dilation, contraction, and consolidation) and shearing behavior were all connected via the theory of plasticity using critical state soil mechanics by Roscoe, Schofield, and Wroth with the publication of "On the Yielding of Soils" in 1958. Critical state soil mechanics is the basis for many contemporary advanced constitutive models describing the behavior of soil.

Geotechnical centrifuge modeling is a method of testing physical scale models of geotechnical problems. The use of a centrifuge enhances the similarity of the scale model tests involving soil because the strength and stiffness of soil is very sensitive to the confining pressure. The centrifugal acceleration allows a researcher to obtain large (prototype-scale) stresses in small physical models.

## Practicing Engineers

Geotechnical engineers are typically graduates of a four-year civil engineering program and some hold a masters degree. In the USA, geotechnical engineers are typically licensed and regulated as Professional Engineers (PEs) in most states; currently only California and Oregon have licensed geotechnical engineering specialties. The Academy of Geo-Professionals (AGP) began issuing Diplomate, Geotechnical Engineering (D.GE) certification in 2008. State governments will typically license engineers who have graduated from an ABET accredited school, passed the Fundamentals of Engineering examination, completed several years of work experience under the supervision of a licensed Professional Engineer, and passed the Professional Engineering examination.

## Weight- Volume Relationship

Basic Definitions

Figure (a) shows a soil mass that has a total volume V and a total weight, W. to develop the weightvolume relationships, the three phases of the soil mass, i.e., soil solids, air, and water, have been separated in Figure (b).

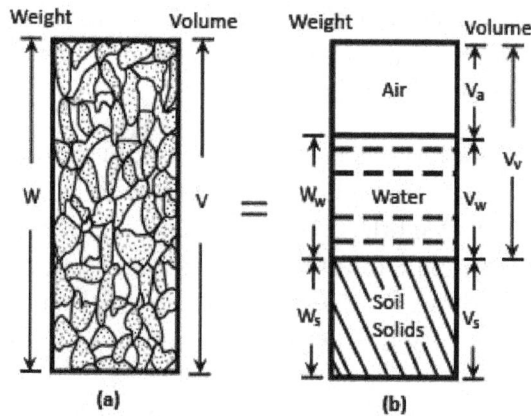

Weight-volume relationships for soil aggregate

$$w = W_s + W_w \qquad (1)$$

And, also,

$$V = V_s + V_w + V_a \qquad (2)$$

$$V_v = V_w + V_a \qquad (3)$$

Where

$$W_s = weight\ of\ soil\ solids$$
$$W_w = weight\ of\ water$$
$$V_s = volume\ of\ the\ soil\ solids$$
$$V_w = volume\ of\ water$$
$$V_a = volume\ of\ air$$

The weight of air is assumed to be zero. The volume relations commonly used in soil mechanics are void ratio, porosity, and degree of saturation.

Void ratio e defined as the ratio of the volume of voids to the volume of solids:

$$e = \frac{V_v}{V_s} \tag{4}$$

Porosity n is defined as the ratio of the volume of voids to the total volume:

$$n = \frac{V_v}{V} \tag{5}$$

Also, $V = V_s + V_v$

And so

$$n = \frac{V_v}{V_s + V_v} = \frac{V_v/V_s}{V_s/V_s + V_v/V_s} = \frac{e}{1+e} \tag{6}$$

Degree of saturation $S_r$ is the ratio of the volume of water to the volume of voids and is generally expressed as a percentage:

$$S_r(\%) = \frac{V_w}{V_v} \times 100 \tag{7}$$

The weight relations used are moisture content and unit weight. Moisture content w is defined as the ratio of the weight of water to the weight of soil solids, generally expressed as a percentage:

$$w(\%) = \frac{W_w}{W_s} \times 100 \tag{8}$$

Unit weight $\gamma$ is the ratio of the total weight to the total volume of the soil aggregate:

$$\gamma = \frac{W}{V} \tag{9}$$

This is sometimes referred to as moist unit weight since it includes the weight of water and the soil solids. If the entire void space is filled with water ((i.e., $V_a = 0$), it is a saturated soil; Eq. (9) will then give use the saturated unit weight $\gamma_{sat}$.

The dry unit weight $\gamma_d$ is defined as the ratio of the weight of soil solids to the total volume:

$$\gamma_d = \frac{W_s}{V} \tag{10}$$

Useful weight-volume relations can be developed by considering a soil mass is which the volume of soil solids is unity, as shown in Figure. Since $V_s = 1$ from the definition of void ratio given in Eq. (4) the volume of voids is equal to the void ratio, e. the weight of soil solids can be given by

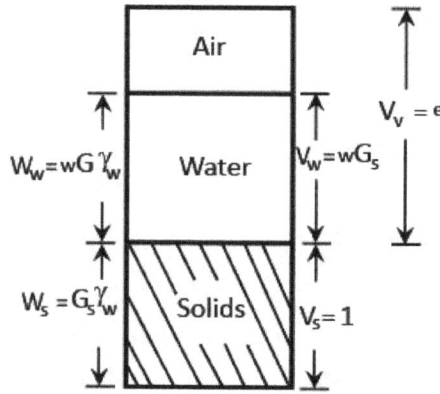

Weight - volume relations for $V_s = 1$

$$W_s = G_s \gamma_w V_s = G_s \gamma_w \quad (since V_s = 1)$$

Where $G_s$ is the specific gravity of soil solids, and $\gamma_w$ is the unit weight of water $(62.4 lb / ft^3$, or $9.81 kN / m^3)$.

From Eq. (8), the weight of water is $W_w = wW_s = wG_s\gamma_w$. So the moist unit weight is

$$\gamma = \frac{W}{V} = \frac{W_s + W_w}{V_s + V_u} = \frac{G_s\gamma_w + wG_s\gamma_w}{1+e} = \frac{G_s\gamma_w(1+w)}{1+e} \tag{11}$$

The dry unit weight can also be determined from Figure as

$$\gamma_d = \frac{W_s}{V} = \frac{G_s\gamma_w}{1+e} \tag{12}$$

The degree of saturation can be given by

$$S_r = \frac{V_w}{V_v} = \frac{W_w / \gamma_w}{V_v} = \frac{wG_s\gamma_w / \gamma_w}{e} = \frac{wG_s}{e} \tag{13}$$

For saturated soils, $S_r = 1$. So, from Eq. (13)

$$e = wG_s \qquad (14)$$

By referring to Figure, the relation for the unit weight of a saturated soil can be obtained as

$$\gamma_{sat} = \frac{W}{V} = \frac{W_s + W_w}{V} = \frac{G_s\gamma_w + e\gamma_w}{1+e} \qquad (15)$$

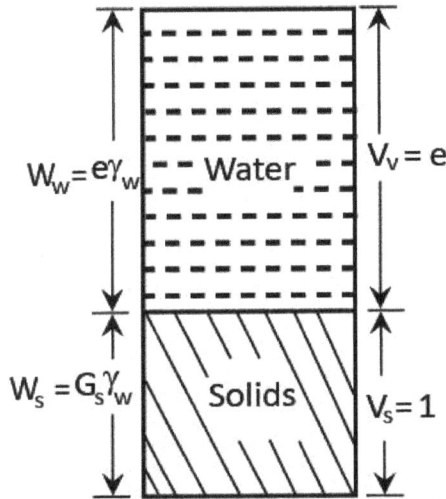

Weight-volume relation for saturated soil with $V_s = 1$

Basic relations for unit weight such as Eqs. (11), (12), and (15) in terms of porosity $n$ can also be derived by considering a soil mass that has a total volume of unity as shown in Figure. In this case (for $V=1$), from Eq. (5) $V_u = n$, So, $V_s = V - V_u = 1 - n$.

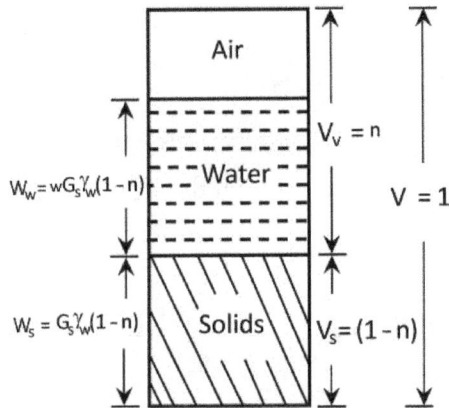

Weight-volume relation with $V = 1$

The weight of soil solids is equal to $(1-n)G_s\gamma_w$, and the weight of water $W_w = wW_s = w(1-n)G_s\gamma_w$. Thus the moist unit weight is

$$\gamma = \frac{W}{V} = \frac{W_s + W_w}{V} = \frac{(1-n)G_s\gamma_w + w(1-n)G_s\gamma_w}{1}$$

$$= G_s\gamma_w(1-n)(1+w) \tag{16}$$

The dry unit weight is

$$\gamma_d = \frac{w_s}{V} = (1-n)\,G_s\,\gamma_w \tag{17}$$

If the soil is saturated

$$\gamma_{sat} = \frac{W_s + W_w}{V} = (1-n)G_s\gamma_w + n\gamma_w = [G_s - n(G_s - 1)]\gamma_w \tag{18}$$

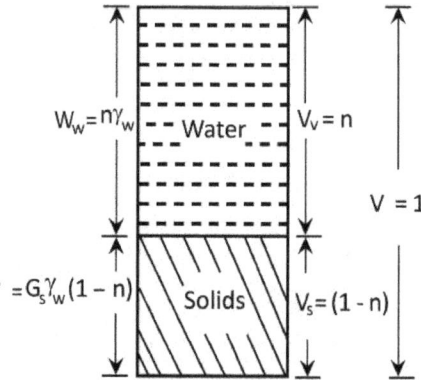

Weight-volume relationship for saturated soil with $V = 1$

From Eq. (12), the dry unit weight is

$$\gamma_d = \frac{G_s\gamma_w}{1+e} = \frac{(2.68)(9.81)}{1+0.8} = 14.61 kN/m^3$$

From Eq. (13), the degree of saturation is

$$S_r(\%) = \frac{wG_s}{e} \times 100 = \frac{(0.24)(2.68)}{0.8} \times 100 = 80.40\%$$

Part (b):From Eq. (14), for saturated soils,

$$e = wG_s, or\ w(\%) = \frac{e}{G_s} \times 100 = \frac{0.8}{2.68} \times 100 = 29.85\%$$

From Eq. (15), the saturated unit weight is

$$\gamma_{sat} = \frac{G_s\gamma_w + e\gamma_w}{1+e} = \frac{9.81(2.68 + 0.8)}{1+0.8} = 18.97 kN/m^3$$

## General Range of Void Ratio and Dry Unit Weight Encountered in Granular Soils

The loosest and the densest possible arrangements that we can obtain from these equal spheres are, respectively, the simple cubic and pyramidal type of packing as shown in Figure. The void corresponding to the simple cubic type of arrangement is 0.91; that for the pyramidal type of arrangement is 0.34. In the case of natural granular soils, particles are neither of equal size nor perfect spheres. The smallsized particles may occupy void spaces between the larger ones, which will tend to reduce the void ratio of natural soils are compared to that for equal spheres.

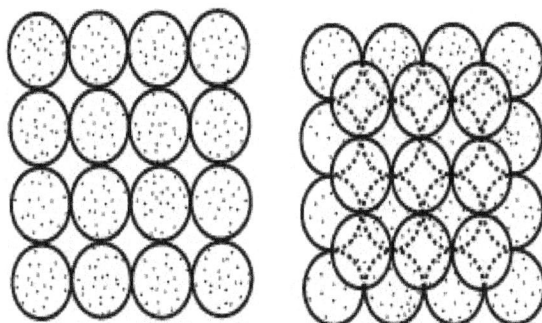

Simple cubic (a) and pyramid (b) types of arrangement of equal spheres.

Table: Gives some typical values of void ratios and dry unit weights encountered in granular soils. Table typical values of void ratios and dry unit weights for granular soils.

| | Void ratio $e$ | | Dry unit weight $\gamma_d$ | |
| | | | Minimum | Maximum |
| ] Soil type | Maximum | Minimum | $kN/m^3$ | $kN/m^3$ |
|---|---|---|---|---|
| Gravel | 0.6 | 0.3 | 16 | 20 |
| Coarse sand | 0.75 | 0.35 | 15 | 19 |
| Fine sand | 0.85 | 0.4 | 14 | 19 |
| Standard | 0.8 | 0.5 | 14 | 17 |
| Gravelly sand | 0.7 | 0.2 | 15 | 22 |
| Silty sand | 1 | 0.4 | 13 | 19 |
| Silty sand and gravel | 0.85 | 0.15 | 14 | 23 |

## Relative Density and Relative Compaction

Relative density is a term generally used to describe the degree of compaction of coarse-grained soils. Relative density $D_r$ is defined as

$$D_r = \frac{e_{max} - e}{e_{max} - e_{min}} \qquad (19)$$

Where

$$e_{max} = \text{maximum posssible void ratio}$$
$$e_{min} = \text{minimum posssible void ratio}$$
$$e = \text{void ratio in natural state of soil}$$

Equation (19) can also be expressed in terms of dry unit weight of the soil:

$$\gamma_{d(max)} = \frac{G_s \gamma_w}{1 + e_{min}}$$

$$Or \; e_{min} = \frac{G_s \gamma_w}{\gamma_d(max)} - 1 \tag{20}$$

Similarly,

$$e_{max} = \frac{G_s \gamma_w}{\gamma_{d(min)}} - 1 \tag{21}$$

$$And \; e = \frac{G_s \gamma_w}{\gamma_d} - 1 \tag{22}$$

The results of the sieve analysis are plotted in Figure.

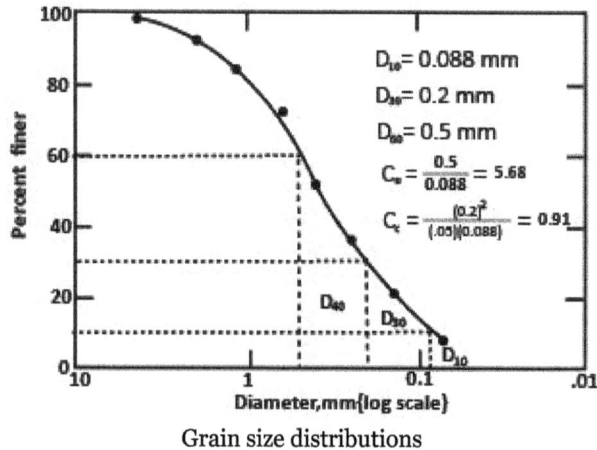

Grain size distributions

The grain-size distribution can be used to determine some of the basic soil parameters such as the effective size, the uniformity coefficient, and the coefficient of gradation. The effective size of a soil is the diameter through which 10% of the total soil mass is passing ad is referred to as $D_{10}$. The uniformity coefficient $C_u$ is defined as

$$C_u = \frac{D_{60}}{D_{10}} \tag{23}$$

Where $D_{60}$ is the diameter through which 60% of the total soil mass is passing. The coefficient of gradation $C_c$ is defined as

$$C_c = \frac{(D_{30})^2}{(D_{60})(D_{10})} \qquad (24)$$

Where $D_{30}$ is the diameter through which 30% of the total soil mass is passing.

The uniformity coefficient and the coefficient of gradation for the sieve analysis shown in Figure.

A soil is called a well-graded soil if the distribution of the grain sizes extends over a rather large range. In that case, the value of the uniformity coefficient is large. Generally, a soil is referred to as well graded if $C_u$ is larger than about 4 to 6 and $C_c$ between 1 and 3. When most of the grains in a soil mass are of approximately the same size – i.e., $C_u$ is close to 1 – the soil is called poorly graded. A soil might have a combination of two or more well-graded soil fractions, and this type of soil is referred to as a gap-graded soil.

For fine-grained soils, the technique used for determination of the grain sizes is hydrometer analysis. This is based on the principle of sedimentation of soil grains. When soil particles are dispersed in water, they will settle at different velocities depending on their weights, shapes, and sizes. For simplicity, it is assumed that all soil particles are spheres, and the velocity of a soil particle can be given by Stokes law as

$$V = \frac{\gamma_s - \gamma_w}{18\eta} D^2 \qquad (25)$$

Where

> $V$ = velocity = distance/ time = $L/t$
>
> $\gamma_w, \gamma_s$ = unit weight of water and soil particles, respectively
>
> $\eta$ = absolute viscosity of water
>
> $D$ = diamter of the soil particles

In the laboratory, hydrometer tests are generally conducted in a sedimentation cylinder, and 50g of ovendried soil is used. The sedimentation cylinder is 18 in (457.2 m) high and 2.5 in (63.5 mm) in diameter, and it is marked for a volume of 1000 ml. a 125-ml solution of 4% sodium hexametaphosphate in distilled water is generally added to the specimen as the dispersing agent. The volume of the dispersed soil suspension is brought up to the 1000 ml mark by adding distilled water. After through mixing, the sedimentation cylinder is placed inside a constant-temperature bath. The hydrometer is then placed in the sedimentation cylinder and readings are taken to the tip of the meniscus at various elapsed times.

When the hydrometer is placed in the soil suspension at a time t after the start of sedimentation, it measures the liquid density in the vicinity of its bulb at a depth $L$. The liquid density is a function of the amount of soil particles present per unit volume of the suspension at that depth. ASTM 152Hydrometers are calibrated to read the amount in grams of soil particles in suspension per 1000 ml (for $G_s = 2.65$ *at a temperature of* $20°C$). Also, at a time $t$ the soil particles in suspension at depth L will have diameters smaller than those calculated by eq. (29), since the larger particles would have settled beyond the zone of measurement. Hence, the percent of soil finer than a given diameter $D$ can be calculated. Since the actual conditions under which the test is conducted may be different from those for which the hydrometers are calibrated ($G_s = 2.65$, *temperature of* $20°C$), it may be necessary to make corrections to the observed hydrometer readings.

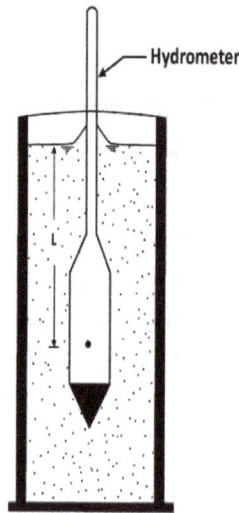

Hydrometer analysis

## Soil Mechanics

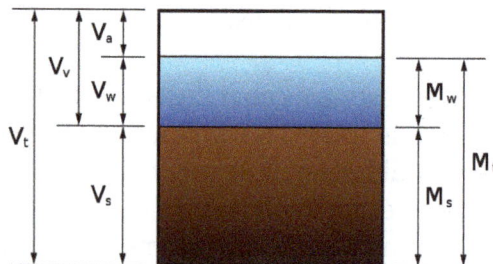

A phase diagram of soil indicating the weights and volumes of air, soil, water, and voids.

In geotechnical engineering, soils are considered a three-phase material composed of: rock or mineral particles, water and air. The voids of a soil, the spaces in between mineral particles, contain the water and air.

The engineering properties of soils are affected by four main factors: the predominant size of the mineral particles, the type of mineral particles, the grain size distribution, and the relative quantities of mineral, water and air present in the soil matrix. Fine particles (fines) are defined as particles less than 0.075 mm in diameter.

## Soil Properties

Some of the important properties of soils that are used by geotechnical engineers to analyze site conditions and design earthworks, retaining structures, and foundations are:

### Specific weight or Unit Weight

Cumulative weight of the solid particles, water and air of the unit volume of soil. Note that the air phase is often assumed to be weightless.

### Porosity

Ratio of the volume of voids (containing air, water, or other fluids) in a soil to the total volume of the soil. Porosity is mathematically related to void ratio the by

$$n = \frac{e}{1+e}$$

here $e$ is void ratio and $n$ is porosity

### Void Ratio

The ratio of the volume of voids to the volume of solid particles in a soil mass. Void ratio is mathematically related to the porosity by

$$e = \frac{n}{1-n}$$

### Permeability

A measure of the ability of water to flow through the soil. It is expressed in units of velocity.

### Compressibility

The rate of change of volume with effective stress. If the pores are filled with water, then the water must be squeezed out of the pores to allow volumetric compression of the soil; this process is called consolidation.

## Shear Strength

The maximum shear stress that can be applied in a soil mass without causing shear failure.

## Atterberg Limits

Liquid limit, Plastic limit, and Shrinkage limit. These indices are used for estimation of other engineering properties and for soil classification.

## Geotechnical Investigation

Geotechnical engineers and engineering geologists perform geotechnical investigations to obtain information on the physical properties of soil and rock underlying (and sometimes adjacent to) a site to design earthworks and foundations for proposed structures, and for repair of distress to earthworks and structures caused by subsurface conditions. A geotechnical investigation will include surface exploration and subsurface exploration of a site. Sometimes, geophysical methods are used to obtain data about sites. Subsurface exploration usually involves in-situ testing (two common examples of in-situ tests are the standard penetration testand cone penetration test). In addition site investigation will often include subsurface sampling and laboratory testing of the soil samples retrieved. The digging of test pits and trenching (particularly for locating faults and slide planes) may also be used to learn about soil conditions at depth. Large diameter borings are rarely used due to safety concerns and expense, but are sometimes used to allow a geologist or engineer to be lowered into the borehole for direct visual and manual examination of the soil and rock stratigraphy.

A variety of soil samplers exist to meet the needs of different engineering projects. The standard penetration test (SPT), which uses a thick-walled split spoon sampler, is the most common way to collect disturbed samples. Piston samplers, employing a thin-walled tube, are most commonly used for the collection of less disturbed samples. More advanced methods, such as ground freezing and the Sherbrooke block sampler, are superior, but even more expensive.

Atterberg limits tests, water content measurements, and grain size analysis, for example, may be performed on disturbed samples obtained from thick walled soil samplers. Properties such as shear strength, stiffness hydraulic conductivity, and coefficient of consolidation may be significantly altered by sample disturbance. To measure these properties in the laboratory, high quality sampling is required. Common tests to measure the strength and stiffness include the triaxial shear and unconfined compression test.

Surface exploration can include geologic mapping, geophysical methods, and photogrammetry; or it can be as simple as an engineer walking around to observe the physical conditions at the site. Geologic mapping and interpretation of geomorphology is typically completed in consultation with a geologist or engineering geologist.

Geophysical exploration is also sometimes used. Geophysical techniques used for subsurface exploration include measurement of seismic waves(pressure, shear, and Rayleigh waves), surface-wave methods and/or downhole methods, and electro-magnetic surveys (magnetometer, resistivity, and ground-penetrating radar).

## Building Foundations

A building's foundation transmits loads from buildings and other structures to the earth. Geotechnical engineers design foundations based on the load characteristics of the structure and the properties of the soils and/or bedrock at the site. In general, ge-otechnical engineers:

1.  Estimate the magnitude and location of the loads to be supported.

2.  Develop an investigation plan to explore the subsurface.

3.  Determine necessary soil parameters through field and lab testing (e.g., consolidation test, triaxial shear test, vane shear test, standard penetration test).

4.  Design the foundation in the safest and most economical manner.

The primary considerations for foundation support are bearing capacity, settlement, and ground movement beneath the foundations. Bearing capacity is the ability of the site soils to support the loads imposed by buildings or structures. Settlement occurs under all foundations in all soil conditions, though lightly loaded structures or rock sites may experience negligible settlements. For heavier structures or softer sites, both overall settlement relative to unbuilt areas or neighboring buildings, and differential settlement under a single structure, can be concerns. Of particular concern is settle-ment which occurs over time, as immediate settlement can usually be compensated for during construction. Ground movement beneath a structure's foundations can occur due to shrinkage or swell of expansive soils due to climatic changes, frost expansion of soil, melting of permafrost, slope instability, or other causes. All these factors must be considered during design of foundations.

Many building codes specify basic foundation design parameters for simple conditions, frequently varying by jurisdiction, but such design techniques are normally limited to certain types of construction and certain types of sites, and are frequently very conser-vative.

In areas of shallow bedrock, most foundations may bear directly on bedrock; in other areas, the soil may provide sufficient strength for the support of structures. In areas of deeper bedrock with soft overlying soils, deep foundations are used to support struc-tures directly on the bedrock; in areas where bedrock is not economically available, stiff "bearing layers" are used to support deep foundations instead.

## Shallow Foundations

Example of a slab-on-grade foundation.

Shallow foundations are a type of foundation that transfers building load to the very near the surface, rather than to a subsurface layer. Shallow foundations typically have a depth to width ratio of less than 1.

## Footings

Footings (often called "spread footings" because they spread the load) are structural elements which transfer structure loads to the ground by direct areal contact. Footings can be isolated footings for point or column loads, or strip footings for wall or other long (line) loads. Footings are normally constructed from reinforced concrete cast directly onto the soil, and are typically embedded into the ground to penetrate through the zone of frost movement and/or to obtain additional bearing capacity.

## Slab Foundations

A variant on spread footings is to have the entire structure bear on a single slab of concrete underlying the entire area of the structure. Slabs must be thick enough to provide sufficient rigidity to spread the bearing loads somewhat uniformly, and to minimize differential settlement across the foundation. In some cases, flexure is allowed and the building is constructed to tolerate small movements of the foundation instead. For small structures, like single-family houses, the slab may be less than 300 mm thick; for larger structures, the foundation slab may be several meters thick.

Slab foundations can be either slab-on-grade foundations or embedded foundations, typically in buildings with basements. Slab-on-grade foundations must be designed to allow for potential ground movement due to changing soil conditions.

## Deep Foundations

Deep foundations are used for structures or heavy loads when shallow foundations cannot provide adequate capacity, due to size and structural limitations. They may also be used to transfer building loads past weak or compressible soil layers. While shallow foundations rely solely on the bearing capacity of the soil beneath them, deep foundations can rely on end bearing resistance, frictional resistance along their length, or both in developing the required capacity. Geotechnical engineers use specialized tools, such as the cone penetration test, to estimate the amount of skin and end bearing resistance available in the subsurface.

Pile-driving for a bridge in Napa, California.

There are many types of deep foundations including piles, drilled shafts, caissons, piers, and earth stabilized columns. Large buildings such as skyscrapers typically require deep foundations. For example, the Jin Mao Tower in China uses tubular steel piles about 1m (3.3 feet) driven to a depth of 83.5m (274 feet) to support its weight.

In buildings that are constructed and found to undergo settlement, underpinning piles can be used to stabilise the existing building.

There are three ways to place piles for a deep foundation. They can be driven, drilled, or installed by use of an auger. Driven piles are extended to their necessary depths with the application of external energy in the same way a nail is hammered. There are four typical hammers used to drive such piles: drop hammers, diesel hammers, hydraulic hammers, and air hammers. Drop hammers simply drop a heavy weight onto the pile to drive it, while diesel hammers use a single cylinder diesel engine to force piles through the Earth. Similarly, hydraulic and air hammers supply energy to piles through hydraulic and air forces. Energy imparted from a hammer head varies with type of hammer chosen, and can be as high as a million foot pounds for large scale diesel hammers, a very common hammer head used in practice. Piles are made of a variety of material including steel, timber, and concrete. Drilled piles are created by first drilling a hole to the appropriate depth, and filling it with concrete. Drilled piles can typically carry more load than driven piles, simply due to a larger diameter pile. The auger method of pile installation is similar to drilled pile installation, but concrete is pumped into the hole as the auger is being removed.

## Lateral Earth Support Structures

A retaining wall is a structure that holds back earth. Retaining walls stabilize soil and rock from downslope movement or erosion and provide support for vertical or near-vertical grade changes. Cofferdams and bulkheads, structures to hold back water, are sometimes also considered retaining walls.

The primary geotechnical concern in design and installation of retaining walls is that the weight of the retained material is creates lateral earth pressure behind the wall, which can cause the wall to deform or fail. The lateral earth pressure depends on the height of the wall, the density of the soil, the strength of the soil, and the amount of allowable movement of the wall. This pressure is smallest at the top and increases toward the bottom in a manner similar to hydraulic pressure, and tends to push the wall away from the backfill. Groundwater behind the wall that is not dissipated by a drainage system causes an additional horizontal hydraulic pressure on the wall.

## Gravity Walls

Gravity walls depend on the size and weight of the wall mass to resist pressures from behind. Gravity walls will often have a slight setback, or batter, to improve wall stability. For short, landscaping walls, gravity walls made from dry-stacked (mortarless) stone or segmental concrete units (masonry units) are commonly used.

Earlier in the 20th century, taller retaining walls were often gravity walls made from large masses of concrete or stone. Today, taller retaining walls are increasingly built as composite gravity walls such as: geosynthetic or steel-reinforced backfill soil with precast facing; gabions (stacked steel wire baskets filled with rocks), crib walls (cells built up log cabin style from precast concrete or timber and filled with soil or free draining gravel) or soil-nailed walls (soil reinforced in place with steel and concrete rods).

For reinforced-soil gravity walls, the soil reinforcement is placed in horizontal layers throughout the height of the wall. Commonly, the soil reinforcement is geogrid, a high-strength polymer mesh, that provide tensile strength to hold soil together. The wall face is often of precast, segmental concrete units that can tolerate some differential movement. The reinforced soil's mass, along with the facing, becomes the gravity wall. The reinforced mass must be built large enough to retain the pressures from the soil behind it. Gravity walls usually must be a minimum of 30 to 40 percent as deep (thick) as the height of the wall, and may have to be larger if there is a slope or surcharge on the wall.

## Cantilever walls

Prior to the introduction of modern reinforced-soil gravity walls, cantilevered walls were the most common type of taller retaining wall. Cantilevered walls are made from a relatively thin stem of steel-reinforced, cast-in-place concrete or mortared masonry (often in the shape of an inverted T). These walls cantilever loads (like a beam) to a large, structural footing; converting horizontal pressures from behind the wall to vertical pressures on the ground below. Sometimes cantilevered walls are buttressed on the front, or include a counterfort on the back, to improve their stability against high loads. Buttresses are short wing wallsat right angles to the main trend of the wall. These walls require rigid concrete footings below seasonal frost depth. This type of wall uses much less material than a traditional gravity wall.

Cantilever walls resist lateral pressures by friction at the base of the wall and/or passive earth pressure, the tendency of the soil to resist lateral movement.

Basements are a form of cantilever walls, but the forces on the basement walls are greater than on conventional walls because the basement wall is not free to move.

## Excavation Shoring

Shoring of temporary excavations frequently requires a wall design which does not extend laterally beyond the wall, so shoring extends below the planned base of the excavation. Common methods of shoring are the use of sheet piles or soldier beams and lagging. Sheet piles are a form of driven piling using thin interlocking sheets of steel to obtain a continuous barrier in the ground, and are driven prior to excavation. Soldier beams are constructed of wide flange steel H sections spaced about 2–3 m apart, driven prior to excavation. As the excavation proceeds, horizontal timber or steel sheeting (lagging) is inserted behind the H pile flanges.

In some cases, the lateral support which can be provided by the shoring wall alone is insufficient to resist the planned lateral loads; in this case additional support is provided by walers or tie-backs. Walers are structural elements which connect across the excavation so that the loads from the soil on either side of the excavation are used to resist each other, or which transfer horizontal loads from the shoring wall to the base of the excavation. Tie-backs are steel tendons drilled into the face of the wall which extend beyond the soil which is applying pressure to the wall, to provide additional lateral resistance to the wall.

## Earthworks

## Excavation

Excavation is the process of training earth according to requirement by removing the soil from the site.

## Filling

Filling is the process of training earth according to requirement by placing the soil on the site.

## Compaction

A compactor/roller operated by U.S. Navy Seabees

Compaction is the process by which the density of soil is increased and permeability of soil is decreased. Fill placement work often has specifications requiring a specific degree of compaction, or alternatively, specific properties of the compacted soil. In-situ soils can be compacted by rolling, deep dynamic compaction, vibration, blasting, gyrating, kneading, compaction grouting etc.

## Ground Improvement

Ground Improvement is a technique that improves the engineering properties of the treated soil mass. Usually, the properties modified are shear strength, stiffness and permeability. Ground improvement has developed into a sophisticated tool to support foundations for a wide variety of structures. Properly applied, i.e. after giving due consideration to the nature of the ground being improved and the type and sensitivity of the structures being built, ground improvement often reduces direct costs and saves time.

## Slope Stabilization

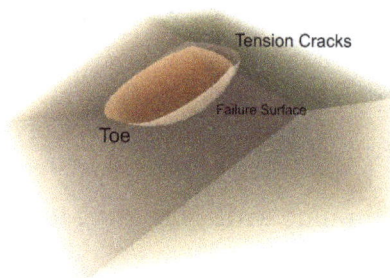

Simple slope slip section.

Slope stability is the potential of soil covered slopes to withstand and undergo movement. Stability is determined by the balance of shear stressand shear strength. A previously stable slope may be initially affected by preparatory factors, making the slope conditionally unstable. Triggering factors of a slope failure can be climatic events can then make a slope actively unstable, leading to mass movements. Mass movements can be caused by increases in shear stress, such as loading, lateral pressure, and transient forces. Alternatively, shear strength may be decreased by weathering, changes in pore water pressure, and organic material.

Several modes of failure for earth slopes include falls, topples, slides, and flows. In slopes with coarse grained soil or rocks, falls typically occur as the rapid descent of rocks and other loose slope material. A slope topples when a large column of soil tilts over its vertical axis at failure. Typical slope stability analysis considers sliding failures, categorized mainly as rotational slides or translational slides. As implied by the name, rotational slides fail along a generally curved surface, while translational slides fail along a more planar surface. A slope failing as a flow would resemble a fluid flowing downhill.

## Slope Stability Analysis

Stability analysis is needed for the design of engineered slopes and for estimating the risk of slope failure in natural or designed slopes. A common assumption is that a slope consists of a layer of soil sitting on top of a rigid base. The mass and the base are assumed to interact via friction. The interface between the mass and the base can be planar, curved, or have some other complex geometry. The goal of a slope stability analysis is to determine the conditions under which the mass will slip relative to the base and lead to slope failure.

If the interface between the mass and the base of a slope has a complex geometry, slope stability analysis is difficult and numerical solution methods are required. Typically, the exact geometry of the interface is not known and a simplified interface geometry is assumed. Finite slopes require three-dimensional models to be analyzed. To keep the problem simple, most slopes are analyzed assuming that the slopes are infinitely wide and can therefore be represented by two-dimensional models. A slope can be drained or undrained. The undrained condition is used in the calculations to produce conservative estimates of risk.

A popular stability analysis approach is based on principles pertaining to the limit equilibrium concept. This method analyzes a finite or infinite slope as if it were about to fail along its sliding failure surface. Equilibrium stresses are calculated along the failure plane, and compared to the soils shear strength as determined by Terzaghi's shear strength equation. Stability is ultimately decided by a factor of safety equal to the ratio of shear strength to the equilibrium stresses along the failure surface. A factor of safety greater than one generally implies a stable slope, failure of which should not occur assuming the slope is undisturbed. A factor of safety of 1.5 for static conditions is commonly used in practice.

## Offshore Geotechnical Engineering

Platforms offshore Mexico.

*Offshore* (or *marine*) *geotechnical engineering* is concerned with foundation design for human-made structures in the sea, away from the coastline (in opposition to *on-shore* or *nearshore*). Oil platforms, artificial islands and submarine pipelines are examples of such structures. There are number of significant differences between onshore and offshore geotechnical engineering. Notably, ground improvement (on the seabed) and site investigation are more expensive, the offshore structures are exposed to a wider range of geohazards, and the environmental and financial consequences are higher in case of failure. Offshore structures are exposed to various environmental loads, notably wind, waves and currents. These phenomena may affect the integrity or the serviceability of the structure and its foundation during its operational lifespan – they need to be taken into account in offshore design.

In subsea geotechnical engineering, seabed materials are considered a two-phase material composed of 1) rock or mineral particles and 2) water. Structures may be fixed in place in the seabed—as is the case for piers, jettys and fixed-bottom wind turbines—or may be a floating structure that remain roughly fixed relative to its geotechnical anchor point. Undersea mooring of human-engineered floating structures include a large number of offshore oil and gas platforms and, since 2008, a few floating wind turbines. Two common types of engineered design for anchoring floating structures include tension-leg and catenary loose mooring systems. "Tension leg mooring systems have vertical tethers under tension providing large restoring moments in pitch and roll. Catenary mooring systems provide station keeping for an offshore structure yet provide little stiffness at low tensions."

## Geosynthetics

Geosynthetics are a type of plastic polymer products used in geotechnical engineering that improve engineering performance while reducing costs. This includes geotextiles, geogrids, geomembranes, geocells, and geocomposites. The synthetic nature of the products make them suitable for use in the ground where high levels of durability are required; their main functions include: drainage, filtration, reinforcement, separation and containment. Geosynthetics are available in a wide range of forms and materials, each to suit a slightly different end use, although they are frequently used together. These products have a wide range of applications and are currently used in many civil and geotechnical engineering applications including: roads, airfields, railroads, embankments, piled embankments, retaining structures, reservoirs, canals, dams, landfills, bank protection and coastal engineering.

A collage of geosynthetic products.

# Soil Science

A sylviculturist, at work

Soil science is the study of soil as a natural resource on the surface of the Earth including soil formation, classification and mapping; physical, chemical, biological, and fertility properties of soils; and these properties in relation to the use and management of soils.

Sometimes terms which refer to branches of soil science, such as pedology (formation, chemistry, morphology and classification of soil) and edaphology (influence of soil on organisms, especially plants), are used as if synonymous with soil science. The diversity of names associated with this discipline is related to the various associations concerned. Indeed, engineers, agronomists, chemists, geologists, physical geographers, ecologists, biologists, microbiologists, silviculturists, sanitarians, archaeologists, and specialists in regional planning, all contribute to further knowledge of soils and the advancement of the soil sciences.

Soil scientists have raised concerns about how to preserve soil and arable land in a world with a growing population, possible future water crisis, increasing per capita food consumption, and land degradation.

## Fields of Study

Soil occupies the pedosphere, one of Earth's spheres that the geosciences use to organize the Earth conceptually. This is the conceptual perspective of pedology and edaphology, the two main branches of soil science. Pedology is the study of soil in its natural setting. Edaphology is the study of soil in relation to soil-dependent uses. Both branches apply a combination of soil physics, soil chemistry, and soil biology. Due to the numerous interactions between the biosphere, atmosphere and hydrosphere that are hosted within the pedosphere, more integrated, less soil-centric concepts are also valuable. Many concepts essential to understanding soil come from individuals not identifiable strictly as soil scientists. This highlights the interdisciplinary nature of soil concepts.

## Research

Dependence on and curiosity about soil, exploring the diversity and dynamics of this resource continues to yield fresh discoveries and insights. New avenues of soil research are compelled by a need to understand soil in the context of climate change, greenhouse gases, and carbon sequestration. Interest in maintaining the planet's biodiversity and in exploring past cultures has also stimulated renewed interest in achieving a more refined understanding of soil.

## Mapping

Most empirical knowledge of soil in nature comes from soil survey efforts. Soil survey, or soil mapping, is the process of determining the soil types or other properties of the soil cover over a landscape, and mapping them for others to understand and use. It relies heavily on distinguishing the individual influences of the five classic soil forming factors. This effort draws upon geomorphology, physical geography, and analysis of vegetation and land-use patterns. Primary data for the soil survey are acquired by field sampling and supported by remote sensing.

## History

Vasily Dokuchaev, a Russian geologist, geographer and early soil scientist, is credited with identifying soil as a resource whose distinctness and complexity deserved to be separated conceptually from geology and crop production and treated as a whole.

Previously, soil had been considered a product of chemical transformations of rocks, a dead substrate from which plants derive nutritious elements. Soil and bedrock were in fact equated. Dokuchaev considers the soil as a natural body having its own genesis and its own history of development, a body with complex and multiform processes taking place within it. The soil is considered as different from bedrock. The latter becomes soil under the influence of a series of soil-formation factors (climate, vegetation, country, relief and age). According to him, soil should be called the "daily" or outward horizons of rocks regardless of the type; they are changed naturally by the common effect of water, air and various kinds of living and dead organisms.

A 1914 encyclopedic definition: "The different forms of earth on the surface of the rocks, formed by the breaking down or weathering of rocks". serves to illustrate the historic view of soil which persisted from the 19th century. Dokuchaev's late 19th century soil concept developed in the 20th century to one of soil as earthy material that has been altered by living processes. A corollary concept is that soil without a living component is simply a part of earth's outer layer.

Further refinement of the soil concept is occurring in view of an appreciation of energy transport and transformation within soil. The term is popularly applied to the material on the surface of the Earth's moon and Mars, a usage acceptable within a portion of the scientific community. Accurate to this modern understanding of soil is Nikiforoff's 1959 definition of soil as the "excited skin of the sub aerial part of the earth's crust".

## Areas of Practice

Academically, soil scientists tend to be drawn to one of five areas of specialization: microbiology, pedology, edaphology, physics or chemistry. Yet the work specifics are very much dictated by the challenges facing our civilization's desire to sustain the land that supports it, and the distinctions between the sub-disciplines of soil science often blur in the process. Soil science professionals commonly stay current in soil chemistry, soil physics, soil microbiology, pedology, and applied soil science in related disciplines.

One interesting effort drawing in soil scientists in the USA as of 2004 is the Soil Quality Initiative. Central to the Soil Quality Initiative is developing indices of soil health and then monitoring them in a way that gives us long term (decade-to-decade) feedback on our performance as stewards of the planet. The effort includes understanding the functions of soil microbiotic crusts and exploring the potential to sequester atmospheric carbon in soil organic matter. The concept of soil quality, however, has not been without its share of controversy and criticism, including critiques by Nobel Laureate Norman Borlaug and World Food Prize Winner Pedro Sanchez.

A more traditional role for soil scientists has been to map soils. Most every area in the United States now has a published soil survey, which includes interpretive tables as to how soil properties support or limit activities and uses. An internationally accepted soil taxonomy allows uniform communication of soil characteristics and functions. National and international soil survey efforts have given the profession unique insights into landscape scale functions. The landscape functions that soil scientists are called upon to address in the field seem to fall roughly into six areas:

- Land-based treatment of wastes
  - Septic system
  - Manure
  - Municipal biosolids
  - Food and fiber processing waste
- Identification and protection of environmentally critical areas
  - Sensitive and unstable soils
  - Wetlands
  - Unique soil situations that support valuable habitat, and ecosystem diversity
- Management for optimum land productivity
  - Silviculture
  - Agronomy
    - Nutrient management
    - Water management
  - Native vegetation
  - Grazing
- Management for optimum water quality
  - Stormwater management
  - Sediment and erosion control
- Remediation and restoration of damaged lands
  - Mine reclamation
  - Flood and storm damage
  - Contamination
- Sustainability of desired uses
  - Soil conservation

There are also practical applications of soil science that might not be apparent from looking at a published soil survey.

- Radiometric dating: specifically a knowledge of local pedology is used to date prior activity at the site

  o Stratification (archeology) where soil formation processes and preservative qualities can inform the study of archaeological sites

  o Geological phenomena

    • Landslides

    • Active faults

- Altering soils to achieve new uses

  o Vitrification to contain radioactive wastes

  o Enhancing soil microbial capabilities in degrading contaminants (bioremediation).

  o Carbon sequestration

  o Environmental soil science

- Pedology

  o Soil genesis

  o Pedometrics

  o Soil morphology

    ▪ Soil micromorphology

  o Soil classification

    ▪ USDA soil taxonomy

- Soil biology

  o Soil microbiology

- Soil chemistry

  o Soil biochemistry

  o Soil mineralogy

- Soil physics

- o   Pedotransfer function
- o   Soil mechanics and engineering
- Soil hydrology, hydropedology

## Fields of Application in Soil Science

- Climate change
- Ecosystem studies
- Pedotransfer function
- Soil fertility / Nutrient management
- Soil management
- Soil survey
- Standard methods of analysis
- Watershed and wetland studies

## Related Disciplines

- Agricultural sciences
  - o   Agricultural soil science
  - o   Agrophysics science
  - o   Irrigation management
- Anthropology
  - o   archaeological stratigraphy
- Environmental science
  - o   Landscape ecology
- Physical geography
  - o   Geomorphology
- Geology
  - o   Biogeochemistry
  - o   Geomicrobiology

- Hydrology
  - Hydrogeology
- Waste management
- Wetland science

# Soil Classification

Soil types

Soil classification deals with the systematic categorization of soils based on distinguishing characteristics as well as criteria that dictate choices in use.

## Overview

Soil classification is a dynamic subject, from the structure of the system itself, to the definitions of classes, and finally in the application in the field. Soil classification can be approached from the perspective of soil as a material and soil as a resource.

## Engineering

Engineers, typically geotechnical engineers, classify soils according to their engineering properties as they relate to use for foundation support or building material. Modern engineering classification systems are designed to allow an easy transition from field observations to basic predictions of soil engineering properties and behaviors.

The most common engineering classification system for soils in North America is the Unified Soil Classification System (USCS). The USCS has three major classification groups: (1) coarse-grained soils (e.g. sands and gravels); (2) fine-grained soils (e.g. silts and clays); and (3) highly organic soils (referred to as "peat"). The USCS further subdivides the three major soil classes for clarification. It distinguishes sands from gravels by grain size, and further classifying some as "well-graded" and the rest as "poorly-graded". Silts and clays are distinguished by the soils' Atterberg limits, and separates "high-plasticity" from "low-plasticity" soils as well. Moderately organic soils are considered subdivisions of silts and clays, and are distinguished from inorganic soils by changes in their plasticity properties (and Atterberg limits) on drying. The European soil classification system (ISO 14688) is very similar, differing primarily in coding and in adding an "intermediate-plasticity" classification for silts and clays, and in minor details.

Other engineering soil classification systems in the United States include the AASHTO Soil Classification System, which classifies soils and aggregates relative to their suitability for pavement construction, and the Modified Burmister system, which works similarly to the USCS, but includes more coding for various soil properties.

A full geotechnical engineering soil description will also include other properties of the soil including color, in-situ moisture content, in-situ strength, and somewhat more detail about the material properties of the soil than is provided by the USCS code. The USCS and additional engineering description is standardized in ASTM D 2487.

## Soil Science

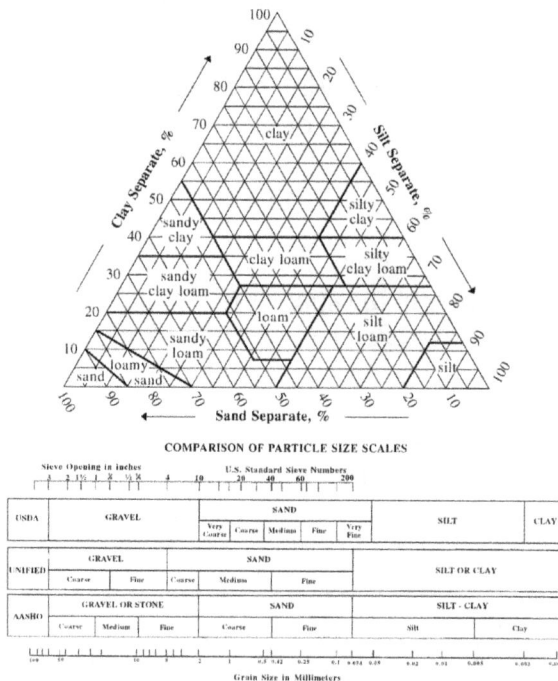

Soil texture triangle showing the USDA classification system based on grain size

For soil resources, experience has shown that a natural system approach to classification, i.e. grouping soils by their intrinsic property (soil morphology), behaviour, or genesis, results in classes that can be interpreted for many diverse uses. Differing concepts of pedogenesis, and differences in the significance of morphological features to various land uses can affect the classification approach. Despite these differences, in a well-constructed system, classification criteria group similar concepts so that interpretations do not vary widely. This is in contrast to a technical system approach to soil classification, where soils are grouped according to their fitness for a specific use and their edaphic characteristics.

Natural system approaches to soil classification, such as the French Soil Reference System are based on presumed soil genesis. Systems have developed, such as USDA soil taxonomy and the World Reference Base for Soil Resources, which use taxonomic criteria involving soil morphology and laboratory tests to inform and refine hierarchical classes.

Another approach is numerical classification, also called ordination, where soil individuals are grouped by multivariate statistical methods such as cluster analysis. This produces natural groupings without requiring any inference about soil genesis.

In soil survey, as practiced in the United States, soil classification usually means criteria based on soil morphology in addition to characteristics developed during soil formation. Criteria are designed to guide choices in land use and soil management. As indicated, this is a hierarchical system that is a hybrid of both *natural* and objective criteria. USDA soil taxonomy provides the core criteria for differentiating soil map units. This is a substantial revision of the 1938 USDA soil taxonomy which was a strictly natural system. Soil taxonomy based soil map units are additionally sorted into classes based on technical classification systems. Land Capability Classes, hydric soil, and prime farmland are some examples.

In addition to scientific soil classification systems, there are also vernacular soil classification systems. Folk taxonomies have been used for millennia, while scientifically based systems are relatively recent developments.

## Soil Classifications for OSHA

The U.S. Occupational Safety and Health Administration (OSHA) requires the classification of soils to protect workers from injury when working in excavations and trenches. OSHA uses 3 soil classifications plus one for rock, based primarily on strength but also:

- Stable Rock: natural solid mineral matter that can be excavated with vertical sides and remain intact while exposed.

- Type A - cohesive, plastic soils with unconfined compressive strength greater than 1.5 ton per square foot (tsf)(144 kPa), and meeting several other requirements (with a lateral soil pressure of 25 psf per ft of depth).

- Type B - cohesive soils with unconfined compressive strength between 0.5 tsf (48 kPa) and 1.5 tsf (144 kPa), or unstable dry rock, or soils which would otherwise be Type A (with a lateral soil pressure of 45 psf per ft of depth).

- Type C - granular soils or cohesive soils with unconfined compressive strength less than 0.5 tsf (48 kPa) or any submerged or freely seeping soil or adversely bedded soils (with a lateral soil pressure of 80 psf per ft of depth).

- Type C60 - A subtype of Type C soil, though is not officially recognized by OSHA as a separate type, has a lateral soil pressure of 60 psf per ft of depth.

Each of the soil classifications has implications for the way the excavation must be made or the protections (sloping, shoring, shielding, etc.) that must be provided to protect workers from collapse of the excavated bank.

## Unified Soil Classification System

The unified system of soil classification was originally proposed by A. Casagrande in 1942 and was then revised in 1952 by the Corps of Engineers and the U.S. Bureau of Reclamation. In its present form, the system s widely used by various organizations, geotechnical engineers in private consulting business, and building codes.

Initially, there are two major divisions in the system. A soil is classified as a coarse-grained soil (gravelly and sandy) if more than 50% is retained on a No. 200 sieve and as a fine-grained soil (silty and clayey) if more than 50% is passing through a No. 200 sieve. The soil is then further classified by a number of subdivisions, as shown in Table. The following symbols are used:

$G$ : *gravel*                     $W$ : *well – graded*

$S$ : *sand*                        $P$ : *poorly graded*

$C$ : *clay*                         $H$ : *high plasticity*

$M$ : *silt*                          $L$ : *low plasticity*

$O$ : *organic silt or clay*

$Pt$ : *peat and highly organic soil*

## Theory of Compaction and Proctor Compaction Test

Compaction of loose fills is a simple way of increasing the stability and load-bearing capacity of soils, and this is generally achieved by using smooth-wheel rollers, sheepsfoot rollers, rubber-tire rollers, and vibratory rollers.

In the compaction process, loose fills are placed in small lifts. Water is then added to the soil to serve as a lubricating agent on the soil particles. With the application of compacting effort, the soil particles slip over each other and move into a densely packed position. The effect of increasing the moisture content is demonstrated in Figure. A silty clay when compacted dry with a compaction effort of $12,375\ ft\,lb\,/\,ft^3$ ($593kJ\,/\,m^3$) can be compacted to a unit weight of $85\,lb\,/\,ft^3$ ($13.4\,kN\,/\,m^3$) However, as the moisture content is increased under the same compactive effort, the weight of soil solids in a unit volume gradually increases. A peak is reached with a moist unit weight of about $125\,lb\,/\,ft^3$ ($19.65kN\,/\,m^3$) at a moisture content of about 20%. So the dry unit weight attained by adding water is

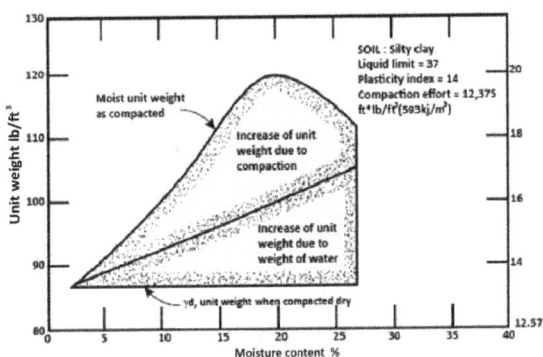

The moisture content vs. unit weight relationship indicating
the increased unit weight resulting from the addition of water and that due
to the compaction effort applied. (Redrawn after A. W. Johnson and J. R. Sallberg,
Factor Influencing Compaction Test Results. Highway Research Board, Bulletin 319, 1962)

$$\gamma d = \frac{\gamma}{1+w} = \frac{125}{1+0.2} = 104.17\,lb\,/\,ft^3\,(16.38kN\,/\,m^3)$$

## Table: unified soil classification system

| Major divisions | Group symbols | Typical names | Criteria of classification * |
|---|---|---|---|
| Course-grained soils (percent passing No. 200 sieve less than 50) | | | |
| Gravels (percent of coarse fraction passing No. 4 sieve less than 50) | | | |
| Gravels with little or no fines | GW | Well-graded gravels, gravel-sand mixtures (little or no fines) | $C_u = \dfrac{D_{60}}{D_{10}} > 4;\ C_c$ $= \dfrac{(D_{10})^2}{D_{10} \times D_{60}}$ between 1 and 3 |
| | GP | Poorly graded gravels, gravel-sand mixtures (little or no fines) | Not meeting the two criteria for GW |
| Gravels with fines | GM | Silty gravels, gravel-sand-silt mixtures | Atterburg limits below "A" line or plasticity index less than 4† |
| | GC | Clayey gravels, gravel-sand-clay mixtures | Atterburg limits above "A" line with plasticity index greater than 7† |
| Sands (percent of coarse fraction passing No. 4 sieve greater than 50) | | | |
| Clean sands (little or no fines) | SW | Well-grraded sands, gravelly sands (little or no fines) | $C_u = \dfrac{D_{60}}{D_{10}} > 6;\ C_c$ $= \dfrac{(D_{10})^2}{D_{10} \times D_{60}}$ between 1 and 3 |
| | SP | Poorly graded sands, gravelly sands (little or no fines) | Not meeting the two criteria for SW |
| Sands with fines (appreciable amount of fines) | SM | Silty sands, sand-silt mixtures | Atterburg limits below "A" line or plasticity index less than 4 † |
| | SC | Clayey sands, sand-clay mixtures | Atterburg limits above "A" line with plasticity index greater than 7† |

| Fine grained soils (percent passing No. 200 sieve greater than 50%) | | | |
|---|---|---|---|
| Silts and clay (liquid limit less than 50) | ML | Inorganic silts, very fine sands, rock flour, silty or clayey fine sands | |
| | CL | Inorganic clays (low to medium plasticity), gravelly clays, sandy clays, silty clays, lean clays | |
| | OL | Organic silts, organic silty clays (low plasticity) | |
| Silts and clay (liquid limit greater than 50) | MH | Inorganic silts. Micaceous or diatomaceous fine sandy or silty soils, elastic silts |  |
| | CH | Inorganic clays (high plasticity), fat clays | |
| Highly organic soils | Pt | Peat, mulch, and other highly organic soils | |

*Classification based on percentage of fines:

| Percent passing No. 200 | classification |
|---|---|
| Less than | 5 GW, GP, SW, SP |
| More than 12 | GM, GC, SM, SC |
| 5 to 12 | borderline-dual symbols required |
| | Such as GW-GM, GW-GC, GP-GM, GP-SC, SW-SM, SW-SC, SP-SM, SP-SC |

† Atterburg limits above "A" line and plasticity index between 4 and 7 are borderline cases. It needs dual symbols.

The effect of compactive effort on the dry unit weight vs. moisture content relation is shown in Figure With increasing compactive effort the optimum moisture content decreases, and the same time the maximum dry unit weight of compaction increases.

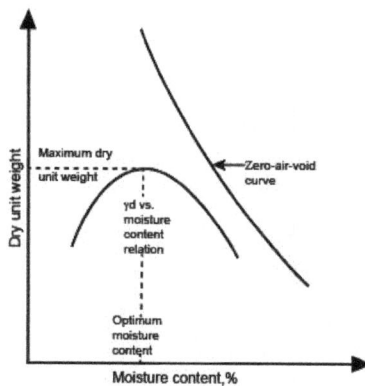

Nature of the variation of dry unit weight of
soil with moisture content in a compaction test

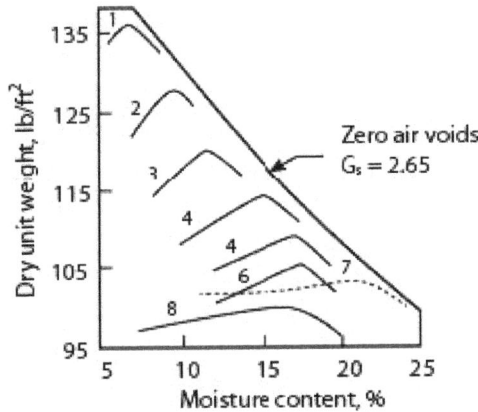

Soil Texture and Plasticity Data

| No. | Description | Sand | Silt | Clay | LL | PI |
|-----|-------------|------|------|------|----|----|
| 1 | Well - graded loamy sand | 88 | 10 | 2 | 16 | NP |
| 2 | Well - graded sandy loam | 78 | 15 | 13 | 16 | NP |
| 3 | Medium - graded sandy loam | 73 | 9 | 18 | 22 | 4 |
| 4 | Lean sandy silty clay | 32 | 33 | 35 | 28 | 9 |
| 5 | Lean silty clay | 5 | 64 | 31 | 36 | 15 |
| 6 | Loessial silt | 5 | 85 | 10 | 26 | 2 |
| 7 | Heavy clay | 6 | 22 | 72 | 67 | 40 |
| 8 | Poorly graded sand | 94 | - 6 - | | NP | - |

Moisture content vs. dry unit weight relationships for eight soils according to AASHTO method T-99. (Note: $1 lb / ft^3 = 157.21 N / m^3$ ). (After A. W. Johnson and J. R. Sallberg, Factors Influencing Compaction Test Results. Highway Research Board, Bulletin 319, 1962)

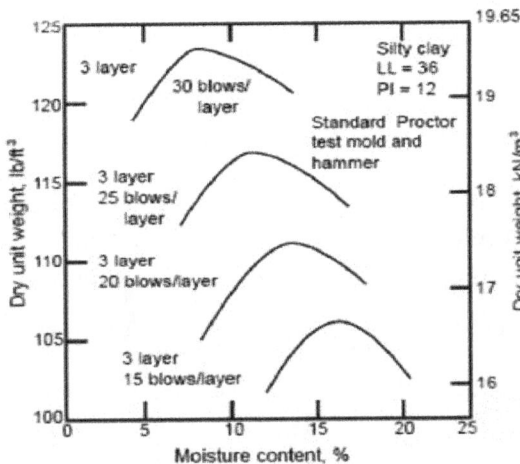

Effect of compactive effort on dry unit weight vs. moisture content relation

With the development of heavier compaction equipment, the standard Proctor test has been modified for better representation of field conditions. In the modified Proctor test (ASTM designation D-1577 and AASHTO designation T-180), the same mold as in the standard Proctor test is used. However, the soil is compacted in five layers with a 10lb (44.5-N) hammer giving 25 bows to each layer. The height of drop of the hammer is 18in (457.2 mm). Hence the compactive effort in the modified Proctor test is equal to

$$\frac{(25\,blows\,/\,\text{layer})(5\,\text{layer})101\,b/\,\text{blow})(1.5-\text{ft drop})}{\dfrac{1}{30}\,ft^3} = 56,250\ ft.lb\,/\,ft^3$$

$(\cong 2694kJ\,/\,m^3)$

Conducting Proctor tests in sandy and a gravelly soil in rather tedious because of lack of control over the moisture content. The nature of the dry unit weight vs. moisture content plot for sand is shown in Figure. With increasing moisture content, the dry unit weight gradually decreases and then increases up to the optimum moisture content. The decrease of dry unit weights obtained at lower moisture contents is a result of the effect of capillary tension in the pore water. The capillary tension resists the movement of soil particles and thus prevents the soil from becoming densely packed.

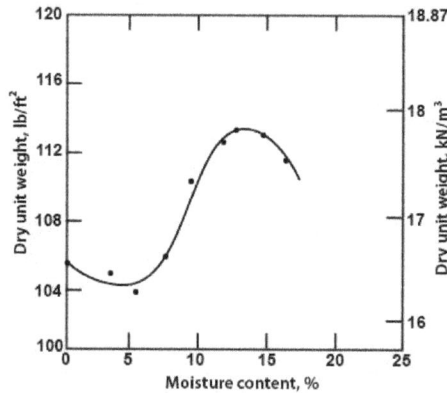

Proctor compation test results on a sand (AASHTO test designation T-99)

## Harvard Miniature Compaction Device

The Harvard miniature compaction device is used in the laboratory for compaction and preparation of soil specimens that are mostly used in research work. Unlike the Proctor test, the compaction is achieved by kneading. The volume of the mold of the Harvard miniature compaction device is $\dfrac{1}{454}\,ft^3\,(62.4cm^3)$. A tamper with a calibrated spring delivers the static pressure to the soil layers. The spring pressure may by 20 lb (89 N) or 40 lb (178 N). The number of layers of soil in the mold and the number of tamps can be varied, thus varying the energy of compaction per unit volume of soil.

## Effect of Organic Content on Compaction of Soil

Soils with high percentage of organic content are often encountered during construction work. Increase of organic content in a soil tends to decrease the maximum dry unit weight of compaction and increase the compressibility of the soil, tendencies which are not desirable in the construction of foundations, embankment, and etc. Franklin et al. (1973) studied the effect of organic contents on the strength and compaction characteristics of mechanical mixtures of inorganic soils and peat and of natural soil samples with the same organic content. The mineralogy of the inorganic fraction of these samples was reasonably the same. Samples for these tests were compacted in the Harvard miniature compaction device with three layers, 40 lb spring force, and 40 tamps per each layer. Figure shows the variation of the maximum dry unit weight of compaction with the organic content, and the variation of the optimum moisture content with the organic content is shown. The organic content O for these soils is defined as

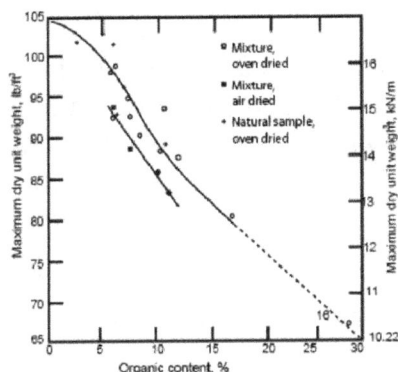

Maximum dry unit weight vs. organic content for all compaction tests. (Note: $1 lb / ft^2 = 157.21 N / m^3$ ). (Redrawn after A. F. Fraklin, L. F. Orozco, and R. Semrau, Compaction of Slightly Organic Soils, J. Soil Mech. Found. Div., ASCE, vol. 99, no. SM7, 1973)

$$O = \frac{loss\ of\ dry\ weight\ due\ to\ heating\ the\ soil\ from\ 105\ to\ 400°C}{dry\ weight\ (at\ 105°C)}$$

Two major conclusions can be drawn from Figures if the organic content in a given soil is more than about 10%, the maximum dry unit weight of compaction decreases considerably. (2) the optimum moisture content increases with the increase of organic contents of soil.

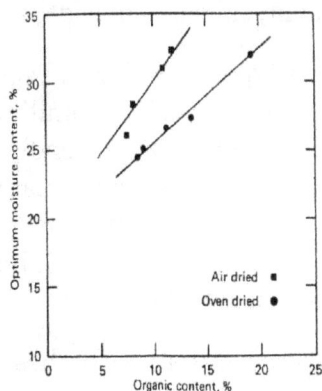

Effect of drying history and organic content on optimum moisture content. (after A. F. Fraklin, L. F. Orozco, and R. Semrau, Compaction of Slightly Organic Soils, J. Soil Mech. Found. Div., ASCE, vol. 99, no. SM7, 1973)

# Soil Stabilization

Soil stabilization a general term for any physical, chemical, biological, or combined method of changing a natural soil to meet an engineering purpose. Improvements include increasing the weight bearing capabilities and performance of in-situ subsoils, sands, and other waste materials in order to strengthen road surfaces.

## General Information

The prime objective of Soil Stabilization is to improve the California Bearing Ratio of in-situ soils by 4 to 6 times. The other prime objective of soil stabilization is to improve on-site materials to create a solid and strong sub-base and base courses. In certain regions of the world, typically developing countries and now more frequently in developed countries, soil stabilization is being used to construct the entire road.

Originally, soil stabilization was done by utilizing the binding properties of clay soils, cement-based products such as soil cement, and/or utilizing the "rammed earth" technique (soil compaction) and lime.

Some of the renewable technologies are: enzymes, surfactants, biopolymers, synthetic polymers, co-polymer based products, cross-linking styrene acrylic polymers, tree resins, ionic stabilizers, fiber reinforcement, calcium chloride, calcite, sodium chloride, magnesium chloride and more. Some of these new stabilizing techniques create hydrophobic surfaces and mass that prevent road failure from water penetration or heavy frosts by inhibiting the ingress of water into the treated layer.

However, recent technology has increased the number of traditional additives used for soil stabilization purposes. Such non-traditional stabilizers include: Polymer based products (e.g. cross-linking water-based styrene acrylic polymers that significantly improves the load-bearing capacity and tensile strength of treated soils), Copolymer Based Products, fiber reinforcement, calcium chloride, and Sodium Chloride.

Traditionally and widely accepted types of soil stabilization techniques use products such as bitumen emulsions which can be used as a binding agents for producing a road base. However, bitumen is not environmentally friendly and becomes brittle when it dries out. Portland cement has been used as an alternative to soil stabilization. However, this can often be expensive and is not a very good "green" alternative. Cement fly ash, lime fly ash (separately, or with cement or lime), bitumen, tar, cement kiln dust (CKD), tree resin and ionic stabilizers are all commonly used stabilizing agents.

There are advantages and disadvantages to many of these soil stabilizers.

Many of the "green" products have essentially the same formula as soap powders, merely lubricating and realigning the soil with no effective binding property. Many of the new approaches rely on large amounts of clay with its inherent binding properties. Bitumen, tar emulsions, asphalt, cement, lime can be used as a binding agents for producing a road base. When using such products issues such as safety, health and the environment must be considered.

The National Society of Professional Engineers (NSPE) has explored some of the newer types of soil stabilization technology, specifically looking for "effective and green" alternatives. One of the examples utilizes new soil stabilization technology, a process based on cross-linking styrene acrylic polymer. Another example uses long crystals to create a closed cell formation that is impermeable to water and is frost, acid, and salt resistant.

Utilizing new soil stabilization technology, a process of cross-linking within the polymeric formulation can replace traditional road/house construction methods in an environmentally friendly and effective way.

There is another soil stabilization method called the Deep Mixing method that is non-destructive and effective at improving load bearing capacity of weak or loose soil strata. This method uses a small, penny-sized injection probe and minimizes debris. This method is ideal for re-compaction and consolidation of weak soil strata, increasing and improving load bearing capacity under structures and the remediation of shallow and deep sinkhole problems. This is particular efficient when there is a need to support deficient public and private infrastructure.

# Soil Compaction

In geotechnical engineering, soil compaction is the process in which a stress applied to a soil causes densification as air is displaced from the pores between the soil grains. When stress is applied that causes densification due to water (or other liquid) being displaced from between the soil grains, then consolidation, not compaction, has occurred. Normally, compaction is the result of heavy machinery compressing the soil, but it can also occur due to the passage of (e.g.) animal feet.

A crawler-backhoe is here equipped with a narrow sheepsfoot roller to compact the fill over newly placed sewer pipe, forming a stable support for a new road surface.

In soil science and agronomy, soil compaction is usually a combination of both engineering compaction and consolidation, so may occur due to a lack of water in the soil, the applied stress being internal suction due to water evaporation as well as due to passage of animal feet. Affected soils become less able to absorb rainfall, thus increasing runoff and erosion. Plants have difficulty in compacted soil because the mineral grains are pressed together, leaving little space for air and water, which are essential for root growth. Burrowing animals also find it a hostile environment, because the denser soil is more difficult to penetrate. The ability of a soil to recover from this type of compaction depends on climate, mineralogy and fauna. Soils with high shrink-swell capacity, such as vertisols, recover quickly from compaction where moisture conditions are variable (dry spells shrink the soil, causing it to crack). But clays which do not crack as they dry cannot recover from compaction on their own unless they host ground-dwelling animals such as earthworms — the Cecil soil series is an example.

A Hamm vibrating roller with plain drum as used
for compacting asphalt and granular soils.

Wacker Neuson vibratory rammer BS 60-2i in action.

## In Construction

Soil compaction is a vital part of the construction process. It is used for support of structural entities such as building foundations, roadways, walkways, and earth retaining structures to name a few. For a given soil type certain properties may deem it more or less desirable to perform adequately for a particular circumstance. In general, the preselected soil should have adequate strength, be relatively incompressible so that future settlement is not significant, be stable against volume change as water content or other factors vary, be durable and safe against deterioration, and possess proper permeability.

When an area is to be filled or backfilled the soil is placed in layers called lifts. The ability of the first fill layers to be properly compacted will depend on the condition of the natural material being covered. If unsuitable material is left in place and backfilled, it may compress over a long period under the weight of the earth fill, causing settlement cracks in the fill or in any structure supported by the fill. In order to determine if the natural soil will support the first fill layers, an area can be proofrolled. Proofrolling consists of utilizing a piece heavy construction equipment (typically, heavy compaction equipment or hauling equipment) to roll across the fill site and watching for deflections to be revealed. These areas will be indicated by the development of rutting, pumping, or ground weaving.

To ensure adequate soil compaction is achieved, project specifications will indicate the required soil density or degree of compaction that must be achieved. These specifications are generally recommended by a geotechnical engineer in a geotechnical engineering report.

The soil type - that is, grain-size distributions, shape of the soil grains, specific gravity of soil solids, and amount and type of clay minerals, present - has a great influence on the maximum dry unit weight and optimum moisture content. It also has a great influence on how the materials should be compacted in given situations. Compaction is accomplished by use of heavy equipment. In sands and gravels, the equipment usually vibrates, to cause re-orientation of the soil particles into a denser configuration. In silts and clays, a sheepsfoot roller is frequently used, to create small zones of intense shearing, which drives air out of the soil.

Determination of adequate compaction is done by determining the in-situ density of the soil and comparing it to the maximum density determined by a laboratory test. The most commonly used laboratory test is called the Proctor compaction test and there are two different methods in obtaining the maximum density. They are the standard Proctor and modified Proctor tests; the modified Proctor is more commonly used. For small dams, the standard Proctor may still be the reference.

While soil under structures and pavements needs to be compacted, it is important after construction to decompact areas to be landscaped so that vegetation can grow.

## Compaction Methods

There are several means of achieving compaction of a material. Some are more appropriate for soil compaction than others, while some techniques are only suitable for particular soils or soils in particular conditions. Some are more suited to compaction of non-soil materials such as asphalt. Generally, those that can apply significant amounts of shear as well as compressive stress, are most effective.

The available techniques can be classified as:

1.  Static - a large stress is slowly applied to the soil and then released.

2.  Impact - the stress is applied by dropping a large mass onto the surface of the soil.

3.  Vibrating - a stress is applied repeatedly and rapidly via a mechanically driven plate or hammer. Often combined with rolling compaction.

4.  Gyrating - a static stress is applied and maintained in one direction while the soil is a subjected to a gyratory motion about the axis of static loading. Limited to laboratory applications.

5.  Rolling - a heavy cylinder is rolled over the surface of the soil. Commonly used on sports pitches. Roller-compactors are often fitted with vibratory devices to enhance their effectiveness.

6.  Kneading - shear is applied by alternating movement in adjacent positions. An example, combined with rolling compaction, is the 'sheepsfoot' roller used in waste compaction at landfills.

The construction plant available to achieve compaction is extremely varied and is described elsewhere.

## Test Methods in Laboratory

Soil compactors are used to perform test methods which cover laboratory compaction methods used to determine the relationship between molding water content and dry unit weight of soils. Soil placed as engineering fill is compacted to a dense state to obtain satisfactory engineering properties such as, shear strength, compressibility, or permeability. In addition, foundation soils are often compacted to improve their engineering properties. Laboratory compaction tests provide the basis for determining the percent compaction and molding water content needed to achieve the required engineering properties, and for controlling construction to assure that the required compaction and water contents are achieved. Test methods such as EN 13286-2, EN 13286-47, ASTM D698, ASTM D1557, AASHTO T99, AASHTO T180, AASHTO T193, BS 1377:4 provide soil compaction testing procedures.

## References

• Jackson, J. A. (1997). Glossary of Geology (4. ed.). Alexandria, Virginia: American Geological Institute. p 604. ISBN 0-922152-34-9

- H. H. Janzen; et al. (2011). "Global Prospects Rooted in Soil Science". Soil Science Society of America Journal. 75 (1): 1. doi:10.2136/sssaj2009.0216

- Buol, S. W.; Hole, F. D. & McCracken, R. J. (1973). Soil Genesis and Classification (First ed.). Ames, IA: Iowa State University Press. ISBN 978-0-8138-1460-5

- C. C. Nikiforoff (1959). "Reappraisal of the soil: Pedogenesis consists of transactions in matter and energy between the soil and its surroundings". Science. 129 (3343): 186–196. Bibcode:1959Sci...129..186N. PMID 17808687. doi:10.1126/science.129.3343.186

- McCarthy, David F. (2007). Essentials of Soil Mechanics and Foundations. Upper Saddle River, NJ: Pearson Prentice Hall. p. 595. ISBN 0-13-114560-6

- Das, Braja M. (2002). Principles of Geotechnical Engineering. Pacific Grove, CA: Brooks/Cole. p. 105. ISBN 0-534-38742-X

- "Soil classification system of England and Wales". Cranfield University, National Soil Resources Institute. Retrieved 2011-12-22

# Clay Minerals: An Integrated Study

Clay is formed with the help of water and is a major element in the speculative theories regarding the origin of life. They are phyllosilicates and are found as sheet structures. Clay is found on Earth as well as other planets. The major categories of clay are dealt with great details in the chapter.

## Clay Minerals

Oxford Clay (Jurassic) exposed near Weymouth, England.

Clay minerals are hydrous aluminium phyllosilicates, sometimes with variable amounts of iron, magnesium, alkali metals, alkaline earths, and other cations found on or near some planetary surfaces.

Clay minerals form in the presence of water and have been important to life, and many theories of abiogenesis involve them. They are important constituents of soils, and have been useful to humans since ancient times in agriculture and manufacturing.

### Properties

Clays form flat hexagonal sheets similar to the micas. Clay minerals are common weathering products (including weathering of feldspar) and low-temperature hydrothermal alteration products. Clay minerals are very common in soils, in fine-grained sedimentary rocks such as shale, mudstone, and siltstone and in fine-grained metamorphic slate and phyllite.

Clay minerals are usually (but not necessarily) ultrafine-grained (normally considered to be less than 2 micrometres in size on standard particle size classifications) and so may require special analytical techniques for their identification and study. These include x-ray diffraction, electron diffraction methods, various spectroscopic methods such as Mössbauer spectroscopy, infrared spectroscopy, Raman spectroscopy, and SEM-EDS or automated mineralogy processes. These methods can be augmented by polarized light microscopy, a traditional technique establishing fundamental occurrences or petrologic relationships.

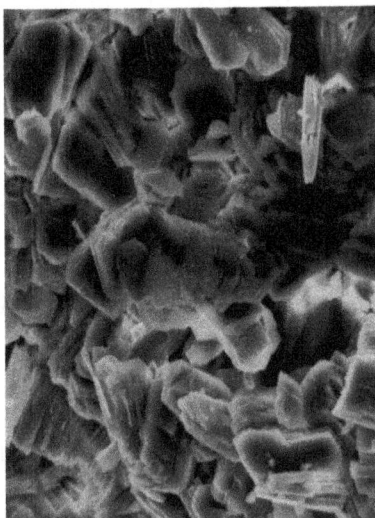

Hexagonal sheets of a clay mineral (kaolinite) (SEM image, x1340 magnification)

## Occurrence

Given the requirement of water, clay minerals are relatively rare in the Solar System, though they occur extensively on Earth where water has interacted with other minerals and organic matter. Clay minerals have been detected at several locations on Mars, including Echus Chasma, Mawrth Vallis, the Memnonia quadrangle and the Elysium quadrangle. Spectrography has confirmed their presence on asteroids including the dwarf planet Ceres and Tempel 1 as well as Jupiter's moon Europa.

## Classification

Clay minerals can be classified as 1:1 or 2:1, this originates because they are fundamentally built of tetrahedral silicate sheets and octahedral hydroxide sheets. A 1:1 clay would consist of one tetrahedral sheet and one octahedral sheet, and examples would be kaolinite and serpentine. A 2:1 clay consists of an octahedral sheet sandwiched between two tetrahedral sheets, and examples are talc, vermiculite and montmorillonite.

Clay minerals include the following groups:

- Kaolin group which includes the minerals kaolinite, dickite, halloysite, and nacrite (polymorphs of $Al_2Si_2O_5(OH)_4$).

- o Some sources include the kaolinite-serpentine group due to structural similarities (Bailey 1980).

- Smectite group which includes dioctahedral smectites such as montmorillonite, nontronite and beidellite and trioctahedral smectites for example saponite. In 2013, analytical tests by the Curiosity rover found results consistent with the presence of *smectite clay minerals* on the planet Mars.

- Illite group which includes the clay-micas. Illite is the only common mineral.

- Chlorite group includes a wide variety of similar minerals with considerable chemical variation.

- Other 2:1 clay types exist such as sepiolite or attapulgite, clays with long water channels internal to their structure.

Mixed layer clay variations exist for most of the above groups. Ordering is described as random or regular ordering, and is further described by the term reichweite, which is German for range or reach. This type would be ordered in an ISISIS fashion. R0 on the other hand describes random ordering, and other advanced ordering types are also found (R3, etc.). Mixed layer clay minerals which are perfect R1 types often get their own names. R1 ordered chlorite-smectite is known as corrensite, R1 illite-smectite is rectorite.

## History

Knowledge of the nature of clay became better understood in the 1930s with advancements in x-ray diffraction technology necessary to analyze the molecular nature of clay particles. Standardization in terminology arose during this period as well with special attention given to similar words that resulted in confusion such as sheet and plane.

## Structure

Like all phyllosilicates, clay minerals are characterised by two-dimensional *sheets* of corner sharing $SiO_4$ tetrahedra and/or $AlO_4$ octahedra. The sheet units have the chemical composition $(Al,Si)_3O_4$. Each silica tetrahedron shares 3 of its vertex oxygen atoms with other tetrahedra forming a hexagonal array in two-dimensions. The fourth vertex is not shared with another tetrahedron and all of the tetrahedra "point" in the same direction; i.e. all of the unshared vertices are on the same side of the sheet.

In clays, the tetrahedral sheets are always bonded to octahedral sheets formed from small cations, such as aluminium or magnesium, and coordinated by six oxygen atoms. The unshared vertex from the tetrahedral sheet also forms part of one side of the octahedral sheet, but an additional oxygen atom is located above the gap in the tetrahedral sheet at the center of the six tetrahedra. This oxygen atom is bonded to a hydrogen atom forming an OH group in the clay structure. Clays can be categorized depending on the way that tetrahedral and octahedral sheets are packaged into *layers*. If there is only one tetrahedral and one octahedral group in each layer the clay is known as a 1:1 clay. The alternative, known as a 2:1 clay, has two tetrahedral sheets with the unshared vertex of each sheet pointing towards each other and forming each side of the octahedral sheet.

Bonding between the tetrahedral and octahedral sheets requires that the tetrahedral sheet becomes corrugated or twisted, causing ditrigonal distortion to the hexagonal array, and the octahedral sheet is flattened. This minimizes the overall bond-valence distortions of the crystallite.

Depending on the composition of the tetrahedral and octahedral sheets, the layer will have no charge, or will have a net negative charge. If the layers are charged this charge is balanced by interlayer cations such as $Na^+$ or $K^+$. In each case the interlayer can also contain water. The crystal structure is formed from a stack of layers interspaced with the interlayers.

## Composition and Structure of Clay Minerals

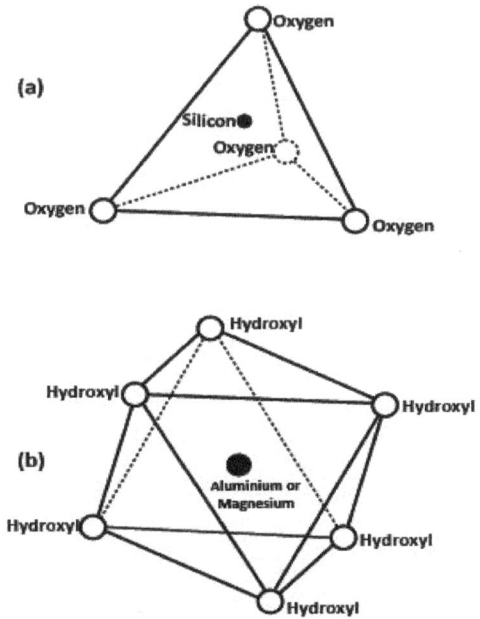

(a) silicon-oxygen tetrahedral unit. (b) Aluminum or magnesium octahedral unit

Clay minerals are complex silicates of aluminum, magnesium, and iron. Two basic crystalline units form the clay minerals: (1) a silicon-oxygen tetrahedron, and (2) an aluminum or magnesium octahedron. A siliconoxygen tetrahedron unit, shown in Figure a, consists of four oxygen atoms surrounding a silicon atom. The tetrahedron units combine to form a silica sheet as shown in Figure. Note that the three oxygen atoms located at the base of each tetrahedron are shared by neighboring tetrahedral. Each silicon atom with a positive valance of 4 is linked to four oxygen atoms with a total negative valance of 8. However, each oxygen atom at the base of the tetrahedron is liked to two silicon atoms. This leaves one negative valance charge of the top oxygen atom of each tetrahedron to be counterbalanced. Figure b shows an octahedral unit consisting of six hydroxyl units surrounding aluminum (or a magnesium) atom. The combination of the aluminum octahedral units forms a gibbsite sheet. If the main metallic atoms in the octahedral units are magnesium, these sheets are referred to as brucite sheets. When the silica sheets are stacked over the octahedral sheets, the oxygen atom replaces the hydroxyls to satisfy their valance bonds. This is shown in Figure c.

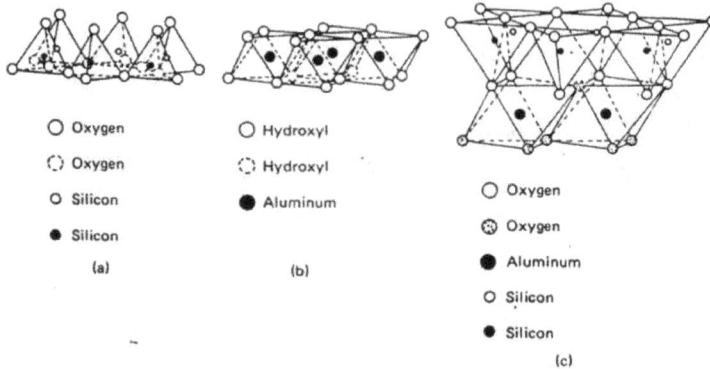

(a) silica sheet. (b) gibbsite sheet. (c) silica-gibbsite sheet. (I: Clay Minerals, J. Soil Mech. Found. Div., ASCE, vol 85 No. SM2 1959.)

Clay minerals with two-layer sheets. Some clay minerals consist of repeating layers of two-layer sheets. A two-layer sheet is a combination of a silica sheet with a gibbsite sheet, or a combination of a silica sheet with a brucite sheet. The sheets are about 7.2Å thick. The repeating layers are held together by hydrogen bonding and secondary valence forces.

Kaolinite is the most important clay mineral belonging to this type. Other common clay mineral that fall into this category are serpentine and halloysite.

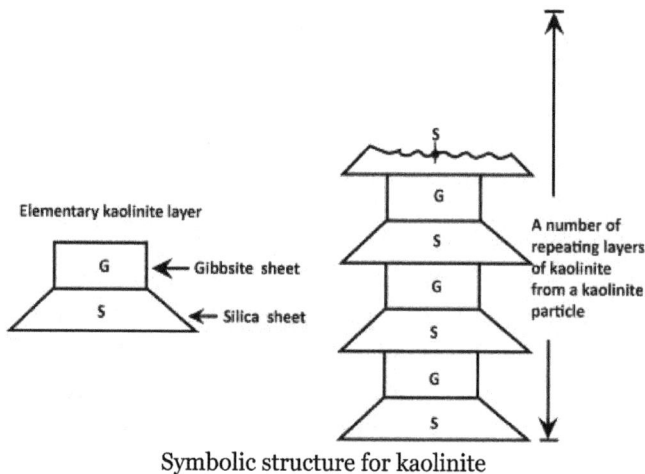

Symbolic structure for kaolinite

## Clay Mineral with Three-layer Sheets

The most common clay mineral with three-layer sheets are illite and montromorillonite. A three-layer sheet consists of an octahedral sheet in the middle with one silica sheet at the top and one at the bottom. Repeated layers of these sheets form the clay minerals.

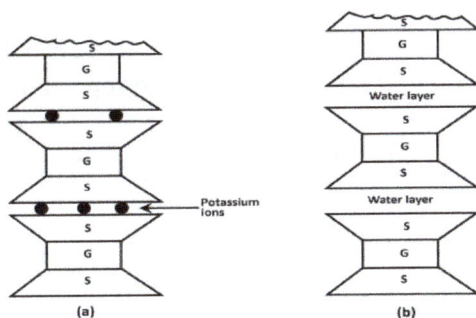

Symbolic structures of (a) illite and (b) montmorillonite

Illite layers are bonded together by potassium ions. The negative charge to balance the potassium ions comes from the substitution of aluminum for some silicon in the tetrahedral sheets. Substitution of this type by one element for another without changing the crystalline form is k now as isomorphous substitution. Montmorillonite has a similar structure to illite. However, unlike illite there are no potassium ions present, and a large amount of water is attracted into the space between the three-sheet layers.

## Cation-exchange Capacity

Cation-exchange capacity (CEC) is a measure of how many cations can be retained on soil particle surfaces. Negative charges on the surfaces of soil particles bind positively-charged atoms or molecules (cations), but allow these to exchange with other cations in the surrounding soil water. This is one of the ways that solid materials in soil alter the chemistry of the soil water. CEC affects many aspects of soil chemistry, and is used as a measure of soil fertility, as it indicates the capacity of the soil to retain several nutrients (e.g. $K^+$, $NH_4^+$, $Ca^{2+}$) in plant-available form. It also indicates the capacity to retain pollutant cations (e.g. $Pb^{2+}$).

## Definition and Principles

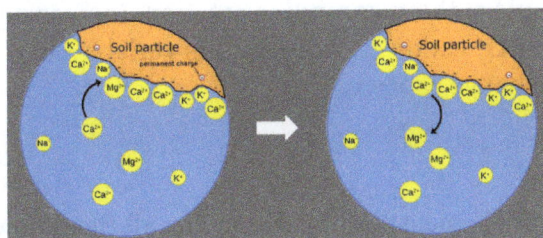

Cation exchange at the surface of a soil particle

Cation-exchange capacity is defined as the amount of positive charge that can be exchanged per mass of soil, usually measured in $cmol_c$/kg. Some texts use the older, equivalent units me/100g or meq/100g. CEC is measured in moles of electric charge, so a cation exchange capacity of 10 $cmol_c$/kg could hold 10 cmol of $Na^+$ cations (with 1 unit of charge per cation) per kilogram of soil, but only 5 cmol $Ca^{2+}$ (2 units of charge per cation).

Cation-exchange capacity arises from various negative charges on soil particle surfaces, especially those of clay minerals and soil organic matter. Phyllosilicate clays consist of layered sheets of aluminium and silicon oxides. The replacement of aluminium or silicon atoms by other elements with lower charge (e.g. $Al^{3+}$ replaced by $Mg^{2+}$) can give the clay structure a net negative charge. This charge does not involve deprotonation and is therefore pH-independent, and called permanent charge. In addition, the edges of these sheets expose many acidic hydroxyl groups that are deprotonated to leave negative charges at the pH levels in many soils. Organic matter also makes a very significant contribution to cation exchange, due to its large number of charged functional groups. CEC is typically higher near the soil surface, where organic matter content is highest, and declines with depth. The CEC of organic matter is highly pH-dependent.

Cations are adsorbed to soil surfaces by the electrostatic interaction between their positive charge and the negative charge of the surface, but they retain a shell of water molecules and do not form direct chemical bonds with the surface.. Exchangeable cations thus form part of the diffuse layer above the charged surface. The binding is relatively weak, and a cation can easily be displaced from the surface by other cations from the surrounding solution.

## pH and CEC

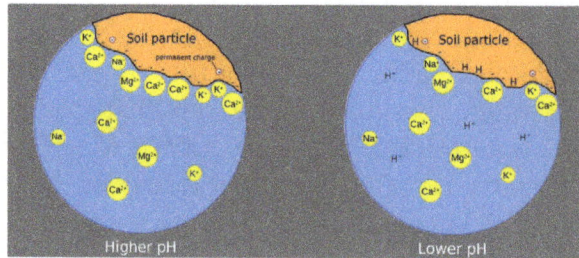

Effect of soil pH on cation exchange capacity

The amount of negative charge from deprotonation of clay hydroxy groups or organic matter depends on the pH of the surrounding solution. Increasing the pH (i.e. decreasing the concentration of $H^+$ cations) increases this variable charge, and therefore also increases the cation exchange capacity.

## Measurement

Cation-exchange capacity is measured by displacing all the bound cations with a concentrated solution of another cation, and then measuring either the displaced cations or the amount of added cation that is retained. Barium ($Ba^{2+}$) and ammonium ($NH_4^+$) are frequently used as exchanger cations, although many other methods are available.

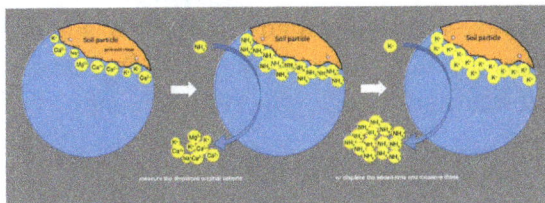

Principle of CEC measurement in soil

CEC measurements depend on pH, and therefore are often made with a buffer solution at a particular pH value. If this pH differs from the natural pH of the soil, the measurement will not reflect the true CEC under normal conditions. Such CEC measurements are called "potential CEC". Alternatively, measurement at the native soil pH is termed "effective CEC", which more closely reflects the real value, but can make direct comparison between soils more difficult.

## Typical Values

The cation exchange capacity of a soil is determined by its constituent materials, which can vary greatly in their individual CEC values. CEC is therefore dependent on parent materials from which the soil developed, and the conditions under which it developed. These factors are also important for determining soil pH, which has a major influence on CEC.

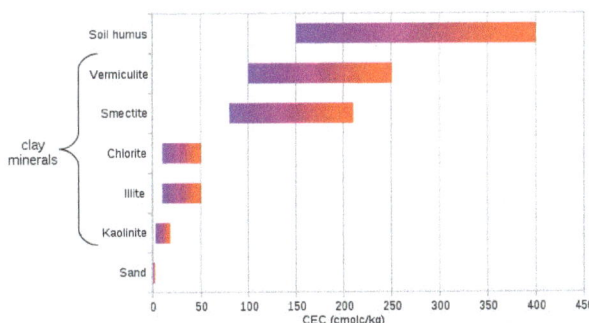

Typical Ranges for CEC of Soil Materials

| Average CEC (pH 7) for some US soils based on USDA Soil Taxonomy | |
|---|---|
| Soil Taxonomy order | CEC ($cmol_c$/kg) |
| Ultisols | 3.5 |
| Alfisols | 9 |
| Spodosols | 9.3 |
| Mollisols | 18.7 |
| Vertisols | 35.6 |
| Entisols | 11.6 |
| Histosols | 128 |

## Base Saturation

Base saturation expresses the percentage of potential CEC occupied by the cations $Ca^{2+}$, $Mg^{2+}$, $K^+$ or $Na^+$. These are traditionally termed "base cations" because they are non-acidic, although they are not bases in the usual chemical sense. Base saturation provides an index of soil weathering and reflects the availability of exchangable cationic nutrients to plants.

## Anion Exchange Capacity

Positive charges of soil minerals can retain anions by the same principle as cation exchange. The surfaces of kaolinite, allophane and iron and aluminium oxides often carry positive charges. In most soils the cation exchange capacity is much greater than the anion exchange capacity, but the opposite can occur in highly weathered soils, such as Ferralsols (Oxisols).

## Nature of Water in Clay

The presence of exchangeable cat ions on the surface of clay particles is familiar to us. Some salt precipitates (cat ions in excess of the exchangeable ions and their associated anions) are also present on the surface of dry clay particles. When water is added to clay, these cat ions and anions float around the clay particles.

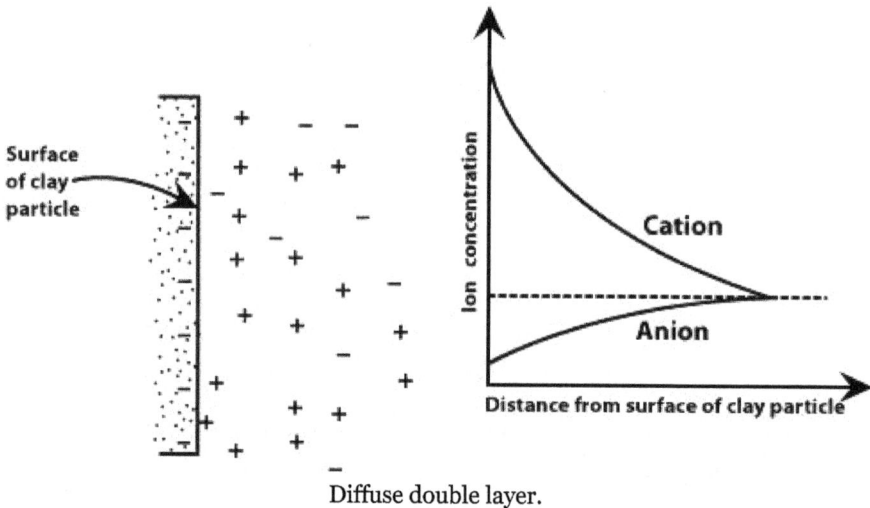

Diffuse double layer.

At this point, it must be pointed out that water molecules are dipolar, since the hydrogen atoms are not symmetrically arrange around the oxygen atoms. This means that a molecule of water is like a rod with positive and negative charges at opposite ends. There are three general mechanisms by which these dipolar water molecules, or dipoles, can be electrically attracted toward the surface of the clay particles.

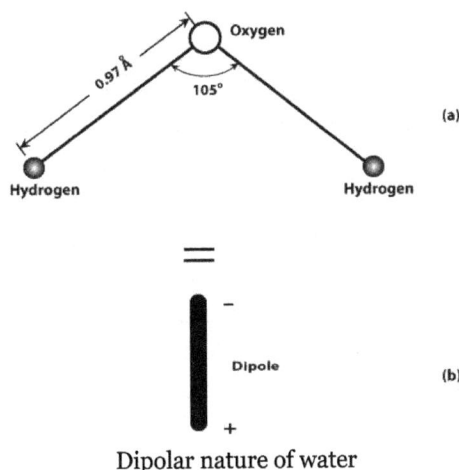

(a)

(b)

Dipolar nature of water

a.  Attraction between the negatively charged faces of clay particles and the positive ends of dipoles.

b.  Attraction between cations in the double layer and the negatively charged ends of dipoles. The cations are in turn attracted by the negatively charged faces of clay particles

c.  Sharing of the hydrogen atoms in the water molecules by hydrogen bonding between the oxygen atoms in the clay particles and the oxygen atoms in the water molecules.

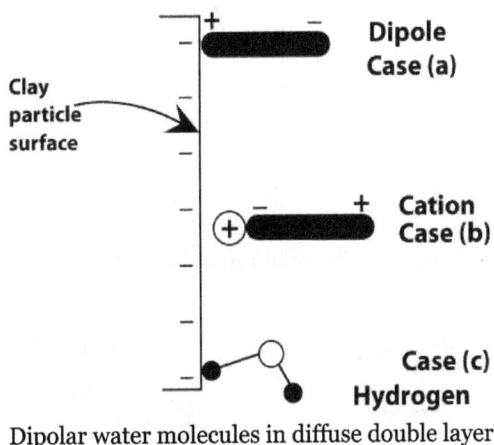

Dipolar water molecules in diffuse double layer

## Flocculation and Dispersion of Clay Particles

In addition to the repulsive force between the clay particles there is an attractive force, which is largely attributed to the Van de Waal's force. This is a secondary bonding force that acts between all adjacent pieces of mater. The force between two flat parallel surfaces varies inversely as $1/x^3$ to $1/x^4$ which x is the distance between the two surfaces. Van der Waal's force is also dependent on the dielectric constant of the medium separating the surfaces. However, if water is the separating medium, substantial changes in the magnitude of the force will not occur with minor changes in the constitution of water.

The behavior of clay particles in a suspension can be qualitatively visualized from our understanding of the attractive and repulsive forces between the particles and with the aid of Figure. Consider a dilute suspension of clay particles in water. These colloidal clay particles will undergo Brownian movement and, during this random movement, will come close to each other at distance within the range of interparticle forces. The forces of attraction and repulsion between the clay particles vary at different rates with respect to the distance of separation. The force of repulsion decreases exponentially with distance, whereas the force of attraction decreases as the inverse third or fourth power of distance, as shown in Figure. Depending on the distance of separation, if the magnitude of the repulsive force is greater than the magnitude of the attractive force, the net result will be repulsion. The clay particles will settle individually and form a dense layer at the bottom; however, they will remain separate from their neighbors. This is referred to as the dispersed state of the soil. On the other hand, if the net force between the particles is attraction, flocs will be formed and these flocs will settle to the bottom. This is called flocculated clay.

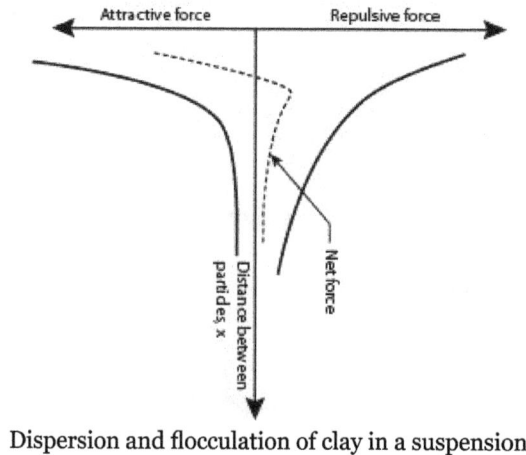

Dispersion and flocculation of clay in a suspension

(a) Dispersion and (b) flocculation of clay

## Salt Flocculation and Non-salt Flocculation

We saw the effect of salt concentration, $n_o$, on the repulsive potential of clay parti-cles. High salt concentration will depress the double layer of clay particles and hence the force of repulsion. We know that the Van der Waal's force largely contributes to the force of attraction between clay particles in suspension. If the clay particles are suspended in water with a high salt concentration, the flocs of the clay particles formed by dominant attractive forces will give them mostly an orientation approaching parallelism (face-to-face type). This is called a salt-type flocculation.

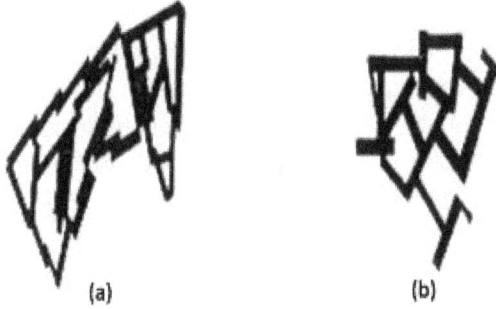

(a)                              (b)

(salt and (b) non-salt flocculation of clay particles. (after T. W. Lamber, Compacted Clay: Structure, Trans. ASCE, vol. 125, 1960

Another type of force of attraction between the clay particles, which is not taken into account in colloidal theories, is that arising the electrostatic attraction of the positive charges at the edge of the particles and the negative charges at the face. In a soil-water suspension with low salt concentration, this electrostatic force attraction may produce a flocculation with an orientation approaching a perpendicular array. This is shown in Figure b and is referred to as non-salt flocculation.

# Expansive Clay

Expansive clay is a type of clay that is known as a lightweight aggregate with a rounded structure, with a porous inner, and a resistant and hard outer layer.

It is a clay or soil that is prone to large volume changes (swelling and shrinking) that are directly related to changes in water content. Soils with a high content of expansive minerals can form deep cracks in drier seasons or years; such soils are called vertisols. Soils with smectite clay minerals, including montmorillonite and bentonite, have the most dramatic shrink-swell capacity.

The mineral make-up of this type of soil is responsible for the moisture retaining capabilities. All clays consist of mineral sheets packaged into layers, and can be classified as either 1:1 or 2:1. These ratios refer to the proportion of tetrahedral sheets to octahedral sheets. Octahedral sheets are sandwiched between two tetrahedral sheets in 2:1 clays, while 1:1 clays have sheets in matched pairs. Expansive clays have an expanding crystal lattice in a 2:1 ratio; however, there are 2:1 non-expansive clays.

Mitigation of the effects of expansive clay on structures built in areas with expansive clays is a major challenge in geotechnical engineering. Some areas mitigate foundation cracking by watering around the foundation with a soaker hose during dry conditions. This process can be automated by a timer, or using a soil moisture sensor controller. Even though irrigation is expensive, the cost is small compared to repairing a cracked foundation. A laboratory test to measure the expansion potential of soil is ASTM D 4829.

Another important characteristic of the expansive clay is its vulnerability to physical changes, according to the amount of water. For example, in a wet season, the clay has the capacity of swelling, and on a dry season, it can shrink and form cracks. According to Biswas and Kriscna, "there are several types of clay minerals of which Montmorillonite has the maximum swelling potential".

## Formation and Characteristics

The expanded clay is obtained by the heating of different types of clay, at a temperature of approximately 1200 °C, using a rotary kiln.

The principal characteristic of the expanded clay is to have a density three times smaller than normal aggregates. Usually, the density of this type of clay is 350 kg/m$^3$.

This kind of clay has a feature that is not very common in lightweight aggregates, since it has a high capacity for thermal and acoustic insulation, which is very important for civil constructions. Also, the use of this material is economically viable.

The expanded clay is very consistent and is more resistant at high temperatures than normal aggregates, and it has higher water absorption. It has a high content of aluminium, silica, oxygen and iron.

The use of the expanded clay aggregate is economically recommended, particularly in the concrete production, since it reduces the bulk density while minimizing the total weight of the constructions. It can be concluded that the use of expanded clay is very important in civil constructions, due to its overall weight, cost and maintenance.

## References

- Napier, W.M.; Wickramasinghe, J.T.; Wickramasinghe, N.C. (2007). "The origin of life in comets". International Journal of Astrobiology. 6 (04). ISSN 1473-5504. doi:10.1017/S1473550407003941

- Agle, DC; Brown, Dwayne (March 12, 2013). "NASA Rover Finds Conditions Once Suited for Ancient Life on Mars". NASA. Retrieved March 12, 2013

- Rivkin, A.S.; Volquardsen, E.L.; Clark, B.E. (2006). "The surface composition of Ceres: Discovery of carbonates and iron-rich clays" (PDF). Icarus. 185 (2): 563–567. doi:10.1016/j.icarus.2006.08.022

- Holmgren, G.G.S.; Meyer, M.W.; Chaney, R.L.; Daniels, R.B. (1993). "Cadmium, Lead, Zinc, Copper, and Nickel in Agricultural Soils of the United States of America". Journal of Environmental Quality. 22: 335–348

- Chang, Kenneth (March 12, 2013). "Mars Could Once Have Supported Life, NASA Says". New York Times. Retrieved March 12, 2013

- Kerr, P.F. (1952). "Formation and Occurrence of Clay Minerals". Clays and Clay Minerals. 1 (1): 19–32. doi:10.1346/CCMN.1952.0010104

# Soil Permeability: An Overview

Soils contain many interconnected pores to provide a route for water to pass. The mean value of the rate of flow is referred to as permeability. This is best understood by Darcy's Law, which provides an equation to calculate rate of flow through pores. The chapter also explores other related concepts such as Navier–Stokes equations, measurement of permeability, etc. The topics discussed in the chapter are of great importance to broaden the existing knowledge on geotechnical engineering and soil science.

## Permeability (Earth Sciences)

Permeability in fluid mechanics and the earth sciences (commonly symbolized as $\kappa$, or $k$) is a measure of the ability of a porous material (often, a rock or an unconsolidated material) to allow fluids to pass through it.

The permeability of a medium is related to the porosity, but also to the shapes of the pores in the medium and their level of connectedness.

### Permeability

Permeability is the property of rocks that is an indication of the ability for fluids (gas or liquid) to flow through rocks. High permeability will allow fluids to move rapidly through rocks. Permeability is affected by the pressure in a rock. The unit of measure is called the darcy, named after Henry Darcy (1803–1858). Sandstones may vary in permeability from less than one to over 50,000 millidarcys (md). Permeabilities are more commonly in the range of tens to hundreds of millidarcies. A rock with 25% porosity and a permeability of 1 md will not yield a significant flow of water. Such "tight" rocks are usually artificially stimulated (fractured or acidized) to create permeability and yield a flow.

### Units

The SI unit for permeability is $m^2$. A practical unit for permeability is the *darcy* (d), or more commonly the *millidarcy* (md) (1 darcy $\approx 10^{-12} m^2$). The name honors the French Engineer Henry Darcy who first described the flow of water through sand filters for potable water supply. Permeability values for sandstones range typically from a fraction of a *darcy* to several *darcys*. The unit of $cm^2$ is also sometimes used (1 $cm^2 = 10^{-4} m^2 \approx 10^8$ d).

## Applications

The concept of permeability is of importance in determining the flow characteristics of hydrocarbons in oil and gas reservoirs, and of groundwater in aquifers.

For a rock to be considered as an exploitable hydrocarbon reservoir without stimulation, its permeability must be greater than approximately 100 md (depending on the nature of the hydrocarbon – gas reservoirs with lower permeabilities are still exploitable because of the lower viscosity of gas with respect to oil). Rocks with permeabilities significantly lower than 100 md can form efficient *seals*. Unconsolidated sands may have permeabilities of over 5000 md.

The concept has also many practical applications outside of geology, for example in chemical engineering (e.g., filtration).

## Description

Permeability is part of the proportionality constant in Darcy's law which relates discharge (flow rate) and fluid physical properties (e.g. viscosity), to a pressure gradient applied to the porous media:

$$v = \frac{\kappa}{\mu}\frac{\Delta P}{\Delta x} \text{ (for linear flow)}$$

Therefore:

$$\kappa = v\frac{\mu\Delta x}{\Delta P}$$

where:

$v$ is the superficial fluid flow velocity through the medium (i.e., the average velocity calculated as if the fluid were the only phase present in the porous medium) (m/s)

$\kappa$ is the permeability of a medium (m$^2$)

$\mu$ is the dynamic viscosity of the fluid (Pa·s)

$\Delta P$ is the applied pressure difference (Pa)

$\Delta x$ is the thickness of the bed of the porous medium (m)

In naturally occurring materials, permeability values range over many orders of magnitude.

## Relation to Hydraulic Conductivity

The proportionality constant specifically for the flow of water through a porous media is called the hydraulic conductivity; permeability is a portion of this, and is a property of the porous media only, not the fluid. Given the value of hydraulic conductivity for a subsurface system, the permeability can be calculated as follows:

$$\kappa = K \frac{\mu}{\rho g}$$

where

- $\kappa$ is the permeability, m$^2$
- $K$ is the hydraulic conductivity, m/s
- $\mu$ is the dynamic viscosity of the fluid, kg/(m·s)
- $\rho$ is the density of the fluid, kg/m$^3$
- $g$ is the acceleration due to gravity, m/s$^2$.

## Determination

Permeability is typically determined in the lab by application of Darcy's law under steady state conditions or, more generally, by application of various solutions to the diffusion equation for unsteady flow conditions.

Permeability needs to be measured, either directly (using Darcy's law), or through estimation using empirically derived formulas. However, for some simple models of porous media, permeability can be calculated (e.g., random close packing of identical spheres).

## Permeability Model based on Conduit Flow

Based on the Hagen–Poiseuille equation for viscous flow in a pipe, permeability can be expressed as:

$$\kappa_I = C \cdot d^2$$

where:

$\kappa_I$ is the intrinsic permeability [length$^2$]

$C$ is a dimensionless constant that is related to the configuration of the flow-paths

$d$ is the average, or effective pore diameter [length].

## Intrinsic and Absolute Permeability

The terms *intrinsic permeability* and *absolute permeability* states that the permeability value in question is an intensive property (not a spatial average of a heterogeneous block of material), that it is a function of the material structure only (and not of the fluid), and explicitly distinguishes the value from that of relative permeability.

## Permeability to Gases

Sometimes permeability to gases can be somewhat different that those for liquids in the same media. One difference is attributable to "slippage" of gas at the interface with the solid when the gas mean free path is comparable to the pore size (about 0.01 to 0.1 µm at standard temperature and pressure). For example, measurement of permeability through sandstones and shales yielded values from $9.0x10^{-19}$ m² to $2.4x10^{-12}$ m² for water and between $1.7x10^{-17}$ m² to $2.6x10^{-12}$ m² for nitrogen gas. Gas permeability of reservoir rock and source rock is important in petroleum engineering, when considering the optimal extraction of shale gas, tight gas, or coalbed methane.

## Tensor Permeability

To model permeability in anisotropic media, a permeability tensor is needed. Pressure can be applied in three directions, and for each direction, permeability can be measured (via Darcy's law in 3D) in three directions, thus leading to a 3 by 3 tensor. The tensor is realised using a 3 by 3 matrix being both symmetric and positive definite (SPD matrix):

- The tensor is symmetric by the Onsager reciprocal relations.

- The tensor is positive definite as the component of the flow parallel to the pressure drop is always in the same direction as the pressure drop.

The permeability tensor is always diagonalizable (being both symmetric and positive definite). The eigenvectors will yield the principal directions of flow, meaning the directions where flow is parallel to the pressure drop, and the eigenvalues representing the principal permeabilities.

## Ranges of Common Intrinsic Permeabilities

These values do not depend on the fluid properties; see the table derived from the same source for values of hydraulic conductivity, which are specific to the material through which the fluid is flowing.

| Permeability | Pervious | | | Semi-Pervious | | | Impervious | | | |
|---|---|---|---|---|---|---|---|---|---|---|
| Unconsolidated sand & gravel | Well sorted gravel | Well sorted sand or sand & gravel | | Very fine sand, silt, loess, loam | | | | | | |
| Unconsolidated clay & organic | | | | Peat | Layered clay | | Unweathered clay | | | |
| Consolidated rocks | Highly fractured rocks | | | Oil reservoir rocks | | Fresh sandstone | Fresh limestone, dolomite | | Fresh granite | |
| $\kappa$ (cm²) | 0.001 | 0.0001 | $10^{-5}$ | $10^{-6}$ | $10^{-7}$ | $10^{-8}$ | $10^{-9}$ | $10^{-10}$ | $10^{-11}$ | $10^{-12}$ | $10^{-13}$ | $10^{-14}$ | $10^{-15}$ |
| $\kappa$ (millidarcy) | $10^{+8}$ | $10^{+7}$ | $10^{+6}$ | $10^{+5}$ | 10,000 | 1,000 | 100 | 10 | 1 | 0.1 | 0.01 | 0.001 | 0.0001 |

## Soil Permeability

The soil mass consists of solid particles of various sizes interconnected void spaces. The continuous void spaces in a soil permit water to flow from a point of high energy to a point of low energy. Permeability is defined as the property of a soil which allows the seepage of fluids through its interconnected void spaces.

## Pore Space in Soil

The pore space of soil contains the liquid and gas phases of soil, i.e., everything but the solid phase that contains mainly minerals of varying sizes as well as organic compounds.

In order to understand porosity better a series of equations have been used to express the quantitative interactions between the three phases of soil.

Macropores or fractures play a major role in infiltration rates in many soils as well as preferential flow patterns, hydraulic conductivity and evapotranspiration. Cracks are also very influential in gas exchange, influencing respiration within soils. Modeling cracks therefore helps understand how these processes work and what the effects of changes in soil cracking such as compaction, can have on these processes.

## Background

## Bulk Density

$$\rho = \frac{M_s}{V_t}$$

The bulk density of soil depends greatly on the mineral make up of soil and the degree of compaction. The density of quartz is around 2.65 g/cm³ but the bulk density of a soil may be less than half that density.

Most soils have a bulk density between 1.0 and 1.6 g/cm³ but organic soil and some friable clay may have a bulk density well below 1 g/cm³.

Core samples are taken by driving a metal core into the earth at the desired depth and soil horizon. The samples are then oven dried and weighed.

Bulk density = (mass of oven dry soil)/volume

The bulk density of soil is inversely related to the porosity of the same soil. The more pore space in a soil the lower the value for bulk density.

## Porosity

$$f = \frac{V_f}{V_t} \quad \text{or} \quad f = \frac{V_a + V_w}{V_s + V_a + V_w}$$

Porosity is a measure of the total pore space in the soil. This is measured as a volume or percent. The amount of porosity in a soil depends on the minerals that make up the soil and the amount of sorting that occurs within the soil structure. For example, a sandy soil will have larger porosity than silty sand, because the silt will fill in the gaps between the sand particles.

## Pore Space Relations

## Hydraulic Conductivity

Hydraulic conductivity (K) is a property of soil that describes the ease with which water can move through pore spaces. It depends on the permeability of the material (pores, compaction) and on the degree of saturation. Saturated hydraulic conductivity, $K_{sat}$, describes water movement through saturated media. Where hydraulic conductivity has the capability to be measured at any state. It can be estimated by numerous kinds of equipment. To calculate hydraulic conductivity, Darcy's law is used. The manipulation of the law depends on the Soil saturation and instrument used.

## Infiltration

Infiltration is the process by which water on the ground surface enters the soil. The water enters the soil through the pores by the forces of gravity and capillary action. The largest cracks and pores offer a great reservoir for the initial flush of water. This allows a rapid infiltration. The smaller pores take longer to fill and rely on capillary forces as well as gravity. The smaller pores have a slower infiltration as the soil becomes more saturated.

## Pore Types

A pore is not simply a void in the solid structure of soil. The various pore size categories have different characteristics and contribute different attributes to soils depending on the number and frequency of each type. A widely used classification of pore size is that of Brewer (1964):

## Macropore

The pores that are too large to have any significant capillary force. Unless impeded, water will drain from these pores, and they are generally air-filled at field capacity. Macropores can be caused by cracking, division of peds and aggregates, as well as plant roots, and zoological exploration. Size >75 μm.

## Mesopore

The largest pores filled with water at field capacity. Also known as storage pores because of the ability to store water useful to plants. They do not have capillary forces too great so that the water does not become limiting to the plants. The properties of mesopores are highly studied by soil scientists because of their impact on agriculture and irrigation. Size 30 µm–75 µm.

## Micropore

These are "pores that are sufficiently small that water within these pores is considered immobile, but available for plant extraction." Because there is little movement of water in these pores, solute movement is mainly by the process of diffusion. Size 5-30 µm.

## Ultramicropore

These pores are suitable for habitation by microorganisms. Their distribution is determined by soil texture and soil organic matter, and they are not greatly affected by compaction Size 0.1-30 µm.

## Cryptopore

Pores that are too small to be penetrated by most microorganisms. Organic matter in these pores is therefore protected from microbial decomposition. They are filled with water unless the soil is very dry, but little of this water is available to plants, and water movement is very slow. Size <0.1 µm.

## Modelling Methods

Basic crack modeling has been undertaken for many years by simple observations and measurements of crack size, distribution, continuity and depth. These observations have either been surface observation or done on profiles in pits. Hand tracing and measurement of crack patterns on paper was one method used prior to advances in modern technology. Another field method was with the use of string and a semicircle of wire. The semi circle was moved along alternating sides of a string line. The cracks within the semicircle were measured for width, length and depth using a ruler. The crack distribution was calculated using the principle of Buffon's needle.

## Disc Permeameter

This method relies on the fact that crack sizes have a range of different water potentials. At zero water potential at the soil surface an estimate of saturated hydraulic conductivity is produced, with all pores filled with water. As the potential is decreased progressively larger cracks drain. By measuring at the hydraulic conductivity at a range of negative potentials, the pore size distribution can be determined. While this is not a physical model of the cracks, it does give an indication to the sizes of pores within the soil.

## Horgan and Young Model

Horgan and Young (2000) produced a computer model to create a two-dimensional prediction of surface crack formation. It used the fact that once cracks come within a certain distance of one another they tend to be attracted to each other. Cracks also tend to turn within a particular range of angles and at some stage a surface aggregate gets to a size that no more cracking will occur. These are often characteristic of a soil and can therefore be measured in the field and used in the model. However it was not able to predict the points at which cracking starts and although random in the formation of crack pattern, in many ways, cracking of soil is often not random, but follows lines of weaknesses.

## Araldite-impregnation Imaging

A large core sample is collected. This is then impregnated with araldite and a fluorescent resin. The core is then cut back using a grinding implement, very gradually (~1 mm per time), and at every interval the surface of the core sample is digitally imaged. The images are then loaded into a computer where they can be analysed. Depth, continuity, surface area and a number of other measurements can then be made on the cracks within the soil.

## Electrical Resistivity Imaging

Using the infinite resistivity of air, the air spaces within a soil can be mapped. A specially designed resistivity meter had improved the meter-soil contact and therefore the area of the reading. This technology can be used to produce images that can be analysed for a range of cracking properties.

## Pore Water Pressure

Pore water pressure (sometimes abbreviated to pwp) refers to the pressure of groundwater held within a soil or rock, in gaps between particles (pores). Pore water pressures in below the phreatic level are measured in piezometers. The vertical pore water pressure distribution in aquifers can generally be assumed to be close to hydrostatic.

In the unsaturated zone, the pore pressure is determined by capillarity and is also referred to as tension, suction, or matric pressure. Pore water pressures under unsaturated conditions (vadose zone) are measured in with tensiometers. Tensiometers operate by allowing the pore water to come into equilibrium with a reference pressure indicator through a permeable ceramic cup placed in contact with the soil.

Pore water pressure is vital in calculating the stress state in the ground soil mechanics, from Terzaghi's expression for the effective stress of a soil.

## General Principles

Pressure develops due to:

- Water elevation difference, water flowing from higher elevation to lower elevation and causing a velocity head, or with water flow, as exemplified in the Bernoulli's energy equations.

- Hydrostatic water pressure, resulting from the weight of material above the point measured.

- Osmotic pressure, inhomogeneous aggregation of ion concentrations, which causes a force in water particles as they attract by the molecular laws of attraction.

- Adsorption pressure, attraction of surrounding soil particles to one another by adsorbed water films.

## Pore Water Pressure below the Water Table

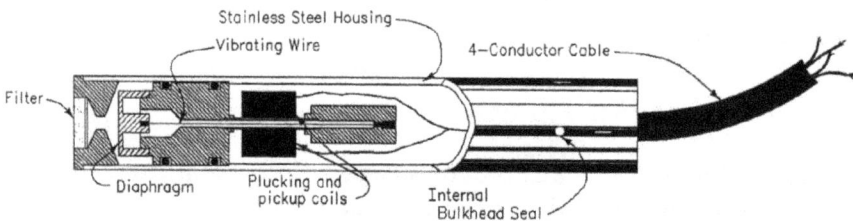

A vibrating wire piezometer. The vibrating wire converts the
fluid pressures into equivalent frequency signals that are then recorded.

The buoyancy effects of water have a large impact on certain soil properties, such as the effective stress present at any point in a soil medium. Consider an arbitrary point five meters below the ground surface. In dry soil, particles at this point experience a total overhead stress equal to the depth underground (5 meters), multiplied by the specific weight of the soil. However, when the local water table height is within said five meters, the total stress felt five meters below surface is decreased by the product of the height of the water table in to the five meter area, and the specific weight of water, 9.81 kN/m^3. This parameter is called the effective stress of the soil, basically equal to the difference in a soil's total stress and pore water pressure. The pore water pressure is essential in differentiating a soil's total stress from its effective stress. A correct representation of stress in soil is necessary for accurate field calculations in a variety of engineering trades.

## Equation for Calculation

When there is no flow, the pore pressure at depth, $h_w$, below the water surface is:

$$p_s = g_w h_w,$$

where:

- $p_s$ is the saturated pore water pressure (kPa),

- $g_w$ is the unit weight of water (kN/m³),
  $g_w = 9.81 kN/m^3$

- $h_w$ is the depth below the water table (m),

## Measurement Methods and Standards

The standard method for measuring pore water pressure below the water table employs a piezometer, which measures the height to which a column of the liquid rises against gravity; i.e., the static pressure (or piezometric head) of groundwater at a specific depth. Piezometers often employ electronic pressure transducers to provide data. The United States Bureau of Reclamation has a standard for monitoring water pressure in a rock mass with piezometers. It sites ASTM D4750, "Standard Test Method for Determining Subsurface Liquid Levels in a Borehole or Monitoring Well (Observation Well)".

## Pore Water Pressure Above the Water Table

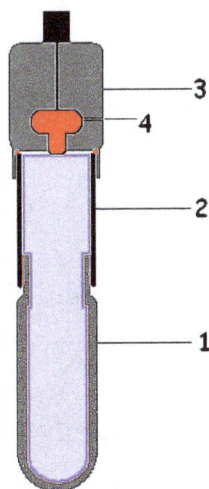

Electronic tensiometer probe: (1) porous cup; (2) water-filled tube; (3) sensor-head; (4) pressure sensor

At any point above the water table, in the vadose zone, the effective stress is approximately equal to the total stress, as proven by Terzaghi's principle. Realistically, the effective stress is greater than the total stress, as the pore water pressure in these partially saturated soils is actually negative. This is primarily due to the surface tension of pore water in voids throughout the vadose zone causing a suction effect on surrounding particles. This capillary action is the "upward movement of water through the vadose zone" (Coduto, 266). Capillary effects in soil are more complex than in free water due to the randomly connected void space and particle interference through which to flow; regardless, the height of this zone of capillary rise, where negative pore water pressure is generally peaks, can be closely approximated by a simple equation. The height of capillary rise is inversely proportional to the diameter of void space in contact with water. Therefore, the smaller the void space, the higher water will rise due to tension forces. Sandy soils consist of more coarse material with more room for voids, and therefore tends to have a much shallower capillary zone than do more cohesive soils, such as clays and silts.

## Equation for calculation

If the water table is at depth $d_w$, then the pore pressure at the ground surface is:

$$p_g = g_w d_w,$$

where:

- $p_g$ is the unsaturated pore water pressure (Pa) at ground level,

- $g_w$ is the unit weight of water (kN/m³),
  $g_w = 9.81 kN / m^3$

- $d_w$ is the depth of the water table (m),

and the pore pressure at depth, z, below the surface is:

$$p_u = g_w (z - d_w),$$

where:

- $p_u$ is the unsaturated pore water pressure (Pa) at point, z, below ground level,

- $z_u$ is depth below ground level.

## Measurement Methods and Standards

A tensiometer is an instrument used to determine the matric water potential ($\Psi_m$) (soil moisture tension) in the vadose zone. An ISO standard, "Soil quality — Determination of pore water pressure — Tensiometer method", ISO 11276:1995, "describes methods for the determination of pore water pressure (point measurements) in unsaturated and saturated soil using tensiometers. Applicable for in situ measurements in the field and, e. g. soil cores, used in experimental examinations." It defines pore water pressure as "the sum of matric and pneumatic pressures".

## Matric Pressure

The amount of work that must be done in order to transport reversibly and isothermally an infinitesimal quantity of water, identical in composition to the soil water, from a pool at the elevation and the external gas pressure of the point under consideration, to the soil water at the point under consideration, divided by the volume of water transported.

## Pneumatic Pressure

The amount of work that must be done in order to transport reversibly and isothermally an infinitesimal quantity of water, identical in composition to the soil water, from a pool at atmospheric pressure and at the elevation of the point under consideration, to a similar pool at an external gas pressure of the point under consideration, divided by the volume of water transported.

# Darcy's Law

Darcy's law is an equation that describes the flow of a fluid through a porous medium. The law was formulated by Henry Darcy based on the results of experiments on the flow of water through beds of sand, forming the basis of hydrogeology, a branch of earth sciences.

## Background

Although Darcy's law (an expression of Newton's second law) was determined experimentally by Darcy, it has since been derived from the Navier–Stokes equations via homogenization. It is analogous to Fourier's law in the field of heat conduction, Ohm's law in the field of electrical networks, or Fick's law in diffusion theory.

One application of Darcy's law is to analyze water flow through an aquifer; Darcy's law along with the equation of conservation of mass are equivalent to the groundwater flow equation, one of the basic relationships of hydrogeology.

Morris Muskat first refined Darcy's equation for single phase flow by including viscosity in the single (fluid) phase equation of Darcy, and this change made it suitable for the petroleum industry. Based on experimental results worked out by his colleagues Wyckoff and Botset, Muskat and Meres also generalized Darcy's law to cover multiphase flow of water, oil and gas in the porous medium of a petroleum reservoir. The generalized multiphase flow equations of Muskat et alios provide the analytical foundation for reservoir engineering that exists to this day.

## Description

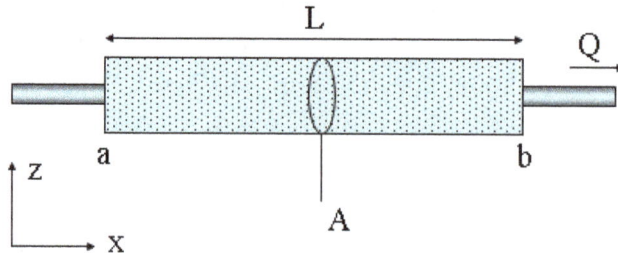

Diagram showing definitions and directions for Darcy's law.

Darcy's law, as refined by Morris Muskat, at constant elevation is a simple proportional relationship between the instantaneous discharge rate through a porous medium, the viscosity of the fluid and the pressure drop over a given distance.

$$Q = -\frac{\kappa A \left( p_b - p_a \right)}{\mu L}.$$

The above equation for single phase (fluid) flow is the defining equation for absolute permeability (single phase permeability). The total discharge, $Q$ (units of volume per time, e.g., m³/s) is equal to the product of the intrinsic permeability of the medium, $\kappa$ (m²), the cross-sectional area to flow, $A$ (units of area, e.g., m²), and the total pressure drop $p_b - p_a$ (pascals), all divided by the viscosity, $\mu$ (Pa·s) and the length over which the pressure drop is taking place ($L$). The negative sign is needed because fluid flows from high pressure to low pressure. Note that the elevation head must be taken into account if the inlet and outlet are at different elevations. If the change in pressure is negative (where $p_a > p_b$), then the flow will be in the positive $x$ direction. There have been several proposals for a constitutive equation for absolute permeability, and the most famous one is probably the Kozeny equation (also called Kozeny–Carman equation).

Dividing both sides of the equation by the area and using more general notation leads

$$q = -\frac{\kappa}{\mu} \nabla p,$$

where $q$ is the flux (discharge per unit area, with units of length per time, m/s) and $\nabla p$ is the pressure gradient vector (Pa/m). This value of flux, often referred to as the Darcy flux or Darcy velocity, is not the velocity which the fluid traveling through the pores is experiencing. The fluid velocity ($v$) is related to the Darcy flux ($q$) by the porosity ($\varphi$). The flux is divided by porosity to account for the fact that only a fraction of the total formation volume is available for flow. The fluid velocity would be the velocity a conservative tracer would experience if carried by the fluid through the formation.

$$v = \frac{q}{\varphi}.$$

Darcy's law is a simple mathematical statement which neatly summarizes several familiar properties that groundwater flowing in aquifers exhibits, including:

- if there is no pressure gradient over a distance, no flow occurs (these are hydrostatic conditions),

- if there is a pressure gradient, flow will occur from high pressure towards low pressure (opposite the direction of increasing gradient — hence the negative sign in Darcy's law),

- the greater the pressure gradient (through the same formation material), the greater the discharge rate, and

- the discharge rate of fluid will often be different — through different formation materials (or even through the same material, in a different direction) — even if the same pressure gradient exists in both cases.

A graphical illustration of the use of the steady-state groundwater flow equation (based on Darcy's law and the conservation of mass) is in the construction of flownets, to quantify the amount of groundwater flowing under a dam.

Darcy's law is only valid for slow, viscous flow; fortunately, most groundwater flow cases fall in this category. Typically any flow with a Reynolds number less than one is clearly laminar, and it would be valid to apply Darcy's law. Experimental tests have shown that flow regimes with Reynolds numbers up to 10 may still be Darcian, as in the case of groundwater flow. The Reynolds number (a dimensionless parameter) for porous media flow is typically expressed as

$$\mathrm{Re} = \frac{\rho v d_{30}}{\mu},$$

where $\rho$ is the density of water (units of mass per volume), $v$ is the specific discharge (not the pore velocity — with units of length per time), $d_{30}$ is a representative grain diameter for the porous media (often taken as the 30% passing size from a grain size analysis using sieves — with units of length), and $\mu$ is the viscosity of the fluid.

## Derivation

For stationary, creeping, incompressible flow, i.e. $\frac{D(\rho u_i)}{Dt} \approx 0$, the Navier–Stokes equation simplifies to the Stokes equation:

$$\mu \nabla^2 u_i + \rho g_i - \partial_i p = 0,$$

where $\mu$ is the viscosity, $u_i$ is the velocity in the $i$ direction, $g_i$ is the gravity component in the $i$ direction and $p$ is the pressure. Assuming the viscous resisting force is linear with the velocity we may write:

$$-\left(\kappa_{ij}\right)^{-1}\mu\varphi u_j + \rho g_i - \partial_i p = 0,$$

where $\varphi$ is the porosity, and $\kappa_{ij}$ is the second order permeability tensor. This gives the velocity in the $n$ direction,

$$\kappa_{ni}\left(\kappa_{ij}\right)^{-1}u_j = \delta_{nj}u_j = u_n = -\frac{\kappa_{ni}}{\varphi\mu}\left(\partial_i p - \rho g_i\right),$$

which gives Darcy's law for the volumetric flux density in the $n$ direction,

$$q_n = -\frac{\kappa_{ni}}{\mu}\left(\partial_i p - \rho g_i\right).$$

In isotropic porous media the off-diagonal elements in the permeability tensor are zero, $\kappa_{ij} = 0$ for $i \neq j$ and the diagonal elements are identical, $\kappa_{ii} = \kappa$, and the common form is obtained

$$q = -\frac{\kappa}{\mu}\left(\nabla p - \rho g\right).$$

The above equation is a governing equation for single phase fluid flow in a porous medium.

## Additional Forms of Darcy's Law

### Darcy's Law in Petroleum Engineering

Another derivation of Darcy's law is used extensively in petroleum engineering to determine the flow through permeable media — the most simple of which is for a one-dimensional, homogeneous rock formation with a single fluid phase and constant fluid viscosity.

$$Q = \frac{\kappa A}{\mu}\left(\frac{\partial p}{\partial x}\right),$$

where $Q$ is the flowrate of the formation (in units of volume per unit time), $k$ is the permeability of the formation (typically in millidarcys), $A$ is the cross-sectional area of the formation, $\mu$ is the viscosity of the fluid (typically in units of centipoise). $\frac{\partial p}{\partial x}$ represents the pressure change per unit length of the formation. This equation can also be solved for permeability and is used to measure it, forcing a fluid of known viscosity through a core of a known length and area, and measuring the pressure drop across the length of the core.

Almost all oil reservoirs have a water zone below the oil leg, and some have also a gas cap above the oil leg. When the reservoir pressure drops due to oil production, water flows into the oil zone from below, and gas flows into the oil zone from above (if the gas cap exists), and we get a simultaneous flow and immiscible mixing of all fluid phases in the oil zone. The operator of the oil field may also inject water (and/or gas) in order to improve oil production. The petroleum industry is therefore using a generalized Darcy equation for multiphase flow that was developed by Muskat et alios. Because Darcy's name is so widespread and strongly associated with flow in porous media, the multiphase equation is denoted Darcy's law for multiphase flow or generalized Darcy equation (or law) or simply Darcy's equation (or law) or simply flow equation if the context says that the text is discussing the multiphase equation of Muskat et alios. Multiphase flow in oil and gas reservoirs is a comprehensive topic, and one of many articles about this topic is Darcy's law for multiphase flow.

## Darcy–Forchheimer Law

For flows in porous media with Reynolds numbers greater than about 1 to 10, inertial effects can also become significant. Sometimes an inertial term is added to the Darcy's equation, known as Forchheimer term. This term is able to account for the non-linear behavior of the pressure difference vs flow data.

$$\frac{\partial p}{\partial x} = -\frac{\mu}{\kappa}q - \frac{\rho}{\kappa_1}q^2,$$

where the additional term $\kappa_1$ is known as inertial permeability.

The flow in the middle of a sandstone reservoir is so slow that Forchheimer's equation is usually not needed, but the gas flow into a gas production well may be high enough to justify use of Forchheimer's equation. In this case the inflow performance calculations for the well, not the grid cell of the 3D model, is based on the Forchheimer equation. The effect of this is that an additional rate-dependent skin appears in the inflow performance formula.

Some carbonate reservoirs have lots of fractures, and Darcy's equation for multiphase flow is generalized in order to govern both flow in fractures and flow in the matrix (i.e. the traditional porous rock). The irregular surface of the fracture walls and high flow rate in the fractures, may justify use of Forchheimer's equation.

## Darcy's Law for Gases in Fine Media (Knudsen Diffusion or Klinkenberg Effect)

For gas flow in small characteristic dimensions (e.g., very fine sand, nanoporous structures etc.), the particle-wall interactions become more frequent, giving rise to additional wall friction (Knudsen friction). For a flow in this region, where both viscous and Knudsen friction are present, a new formulation needs to be used. Knudsen presented a semi-empirical model for flow in transition regime based on his experiments on small capillaries. For a porous medium, the Knudsen equation can be given as

$$N = -\left( \frac{\kappa}{\mu} \frac{p_a + p_b}{2} + D_K^{\text{eff}} \right) \frac{1}{R_g T} \frac{p_b - p_a}{L},$$

where $N$ is the molar flux, $R_g$ is the gas constant, $T$ is the temperature, $D_k^{\text{eff}}$ is the effective Knudsen diffusivity of the porous media. The model can also be derived from the first-principle-based binary friction model (BFM). The differential equation of transition flow in porous media based on BFM is given as

$$\frac{\partial p}{\partial x} = -R_g T \left( \frac{\kappa p}{\mu} + D_K \right)^{-1} N.$$

This equation is valid for capillaries as well as porous media. The terminology of the Knudsen effect and Knudsen diffusivity is more common in mechanical and chemical engineering. In geological and petrochemical engineering, this effect is known as the Klinkenberg effect. Using the definition of molar flux, the above equation can be rewritten as

$$\frac{\partial p}{\partial x} = -R_g T \left( \frac{\kappa p}{\mu} + D_K \right)^{-1} \frac{p}{R_g T} q.$$

This equation can be rearranged into the following equation

$$q = -\frac{\kappa}{\mu} \left( 1 + \frac{D_K \mu}{\kappa} \frac{1}{p} \right) \frac{\partial p}{\partial x}.$$

Comparing this equation with conventional Darcy's law, a new formulation can be given as

$$q = -\frac{\kappa^{\text{eff}}}{\mu} \frac{\partial p}{\partial x},$$

where

$$\kappa^{\text{eff}} = \kappa \left( 1 + \frac{D_K \mu}{\kappa} \frac{1}{p} \right).$$

This is equivalent to the effective permeability formulation proposed by Klinkenberg:

$$\kappa^{\text{eff}} = \kappa \left( 1 + \frac{b}{p} \right).$$

where $b$ is known as the Klinkenberg parameter, which depends on the gas and the porous medium structure. This is quite evident if we compare the above formulations. The Klinkenberg parameter $b$ is dependent on permeability, Knudsen diffusivity and viscosity (i.e., both gas and porous medium properties).

## Darcy's Law for Short Time Scales

For very short time scales, a time derivative of flux may be added to Darcy's law, which results in valid solutions at very small times (in heat transfer, this is called the modified form of Fourier's law),

$$\tau \frac{\partial q}{\partial t} + q = -\kappa \nabla h,$$

where $\tau$ is a very small time constant which causes this equation to reduce to the normal form of Darcy's law at "normal" times (> nanoseconds). The main reason for doing this is that the regular groundwater flow equation (diffusion equation) leads to singularities at constant head boundaries at very small times. This form is more mathematically rigorous, but leads to a hyperbolic groundwater flow equation, which is more difficult to solve and is only useful at very small times, typically out of the realm of practical use.

## Brinkman form of Darcy's Law

Another extension to the traditional form of Darcy's law is the Brinkman term, which is used to account for transitional flow between boundaries (introduced by Brinkman in 1949),

$$\beta \nabla^2 q + q = -\frac{\kappa}{\mu} \nabla p,$$

where $\beta$ is an effective viscosity term. This correction term accounts for flow through medium where the grains of the media are porous themselves, but is difficult to use, and is typically neglected.

## Validity of Darcy's Law

Darcy's law is valid for laminar flow through sediments. In fine-grained sediments, the dimensions of interstices are small and thus flow is laminar. Coarse-grained sediments also behave similarly but in very coarse-grained sediments the flow may be turbulent. Hence Darcy's law is not always valid in such sediments. For flow through commercial pipes, the flow is laminar when Reynolds number is less than 2000, but in some sediments it has been found that flow is laminar when the value of Reynolds number is less than 1.

Considering Figure, the cross-sectional area of the soil is equal to A and the rate of seepage is q.

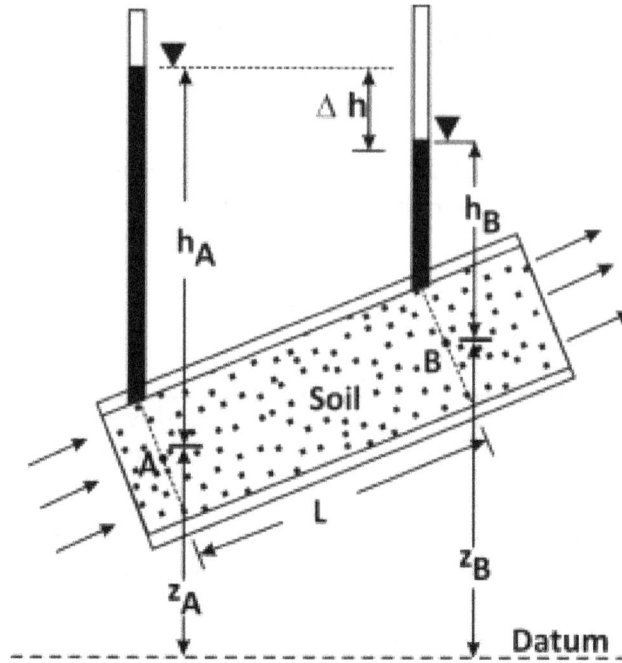

Development of Darcy's law

According to Bernoulli's theorem, the total head for flow at any section in the soil can be given by

Total head = elevation head + pressure head + velocity head          (1)

The velocity head for flow through soil is very small and can be neglected. So, the total heads at section A and B can be given by

Total head at $A = Z_A + h_A$

Total heat at $B = Z_B + h_B$

Where $Z_A$ and $Z_B$ are the elevation heads, and $h_A$ and $h_B$ are the pressure heads.

The loss of head $\Delta h$ between sections A and B is

$$\Delta h = (Z_A + h_A) - (Z_B + Z_B)$$          (2)

The hydraulic gradient $i$ can be written as

$$i = \frac{\Delta h}{L}$$          (3)

Where L is the distance between sections A and B.

Darcy (1856) published a simple relation between the discharge velocity and the hydraulic gradient:

$$v = ki \tag{4}$$

Where

$v$ = Discharge velocity

$i$ = Hydraulic gradient

$k$ = Coefficient of permeability

Hence, the rate of seepage q can be given by

$$q = kiA \tag{5}$$

Note that A is the cross section of the soil perpendicular to the direction of flow.

The coefficient of permeability k has the units of velocity, such as cm/ s or mm/s, and is a measure of the resistance of the soil to flow of water. When the properties of water affecting the flow are included, we can express k by the relation

$$k(cm/s) = \frac{k\rho g}{\mu} \tag{6}$$

Where

$k$ = intrinsic permeability, cm²

$\rho$ = mass density of the fluid, g/ cm³

g = acceleration due to gravity, cm/ sec²

$\mu$ = absolute viscosity of the fluid, poise [that is, g/ (cm. s)]

It must be pointed out that the velocity v given by Equation (4) is the discharge velocity calculated on the basis of the gross cross-sectional area. Since water can flow only through the interconnected pore spaces, the actual velocity of seepage through soil, can be given by

$$V_s = \frac{v}{n} \tag{7}$$

Where n is the porosity of the soil.

Some typical values of the coefficient of permeability are given in table.

The coefficient of permeability of soils is generally expressed at a temperature of 20° C. At any other temperature T, the coefficient of permeability can be obtained from equation. (6) as

$$\frac{k_{20}}{k_T} = \frac{(\rho_{20})(\mu_r)}{(\rho_T)(\mu_{20})}$$

Where

$k_T, k_{20}$ = coefficient of permeability at $T°C$ and $20°C$, respectively

$\rho_T, \rho_{20}$ = mass density of the fluid at $T°C$ and $20°C$, respectively

$\mu_T, \mu_{20}$ = cofficient of viscosity at $T°C$ and $20°C$, respectively

Since the value of $\rho_{20} / \rho_T$ is approximately 1, hence

$$k_{20} = k_T \frac{\mu_T}{\mu_{20}} \qquad (8)$$

Table gives the values of $\mu_T / \mu_{20}$ for a temperature T varying from 10 to 30° C.

Table: Typical values of coefficient of permeability for various soils

| Material | Coefficient of permeability, mm/s |
|---|---|
| Coarse | 10 to $10^3$ |
| Fine gravel, coarse and medium sand | $10^{-2}$ to 10 |
| Fine sand, clayey silt | $10^{-4}$ to $10^{-4}$ |
| Dense silt, clayey silt | $10^{-4}$ to $10^{-4}$ |
| Silty clay, clay | $10^{-5}$ to $10^{-5}$ |

| Temperature T,°C | $\mu_T/\mu_{20}$ | Temperature T,°C | $\mu_T/\mu_{20}$ |
|---|---|---|---|
| 10 | 1.298 | 21 | 0.975 |
| 11 | 1.263 | 22 | 0.952 |
| 12 | 1.228 | 23 | 0.930 |
| 13 | 1.195 | 24 | 0.908 |
| 14 | 1.165 | 25 | 0.887 |
| 15 | 1.135 | 26 | 0.867 |
| 16 | 1.106 | 27 | 0.847 |
| 17 | 1.078 | 28 | 0.829 |
| 18 | 1.051 | 29 | 0.811 |
| 19 | 1.025 | 30 | 0.793 |
| 20 | 1.000 | | |

## Validity of Darcy's Law

Darcy's law given in equation (4), $v = ki$ , is true for laminar flow of water through the void spaces. Several studies have been made to investigate the range over which Darcy's law is valid, and an excellent summary of these works was given by Muskat (1973). A criterion for investigating the range can be furnished by Reynolds number. For flow through soils, Reynolds number    can be given by the relation

$$R_n = \frac{vD\rho}{\mu} \qquad (9)$$

Where

$v =$ discharge (superficial) velocity, cm/s

$D =$ average diameter of the soil particle, cm

$\rho =$ density of the fluid, $g/cm^3$

$\mu =$ coefficient of viscosity, $g/(cm \cdot s)]$

For laminar flow conditions in soils, experimental results show that

$$R_n = \frac{vD\rho}{\mu} \leq 1 \qquad (10)$$

With coarse sand, assuming D = 0.45 mm and making use of equation (7), we have

$k \approx 100D^2 = 100(0.45)^2 = 0.203 cm/s.$ Assuming $i = 1, then\, v = ki = 0.203 cm/s.$

Also, $\rho_{water} \approx 1 g/cm^3$, and $\mu_{20^\circ C} = (10^{-5})(981) g/(cm \cdot) s$.

Hence $R_n = \dfrac{(0.203)(0.045)(1)}{(10^{-5})(981)} = 0.931 < 1$

From the above calculations, we can conclude that, for flow of water through all types of sol (sand, silt, and clay), the flow is laminar and Darcy's law is valid. With coarse sands, gravels, and turbulent flow of water can be expected, and the hydraulic gradient can be given by the relation

$$i = av + bv^2 \qquad (11)$$

Where a and b are experimental constants (Forchheimer 1902).

Leps (1973) summarized a number of works concerned with the determination of the velocity of flow through clean gravel and rocks. All investigators appear to agree that the average velocity of flow through the void spaces can be given by the relation

$$u_v = CR_H^{0.5} i^{0.54} \qquad (12)$$

Where

u $=$ average velocity of flow through voids

C$=$ a constant which is a function of shape and roughness of rock particles

R $=$ hydraulic mean radius

i$=$ hydraulic gradient

# Factor Affecting the Coefficient of Permeability

The coefficient of permeability depends on several factors, most of which are listed below: Shape and size of the soil particles.

- Void ratio : Permeability increases with increase of void ratio.

- Degree of saturation : Permeability increases with increase of degree of saturation. The variation of the value of k with degree of saturation for Madison sand is shown in Figure. Figure show the effect of the degree of saturation on the value of k for a silty clay specimens were prepared by kneading compaction to a dry unit weight of $16.98 kN / m^3$. The molding moisture contents were varied.

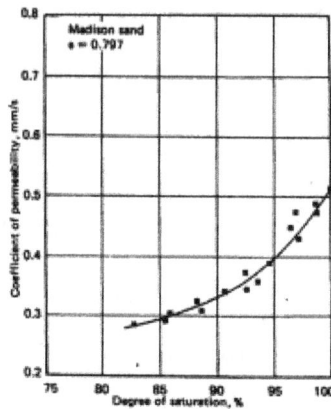

Influence of degree of saturation on permeability of Madison sand

Composition of soil particles: For sands and silts this is not important; however, for soils with clay minerals this is one of the most important factors. Permeability in this case depends on the thickness of water held to the soil particles, which is a function of the cation exchange capacity, valence of the cations, etc. other factors remaining the same, the coefficient of permeability decreases with increasing thickness of the diffuse double layer.

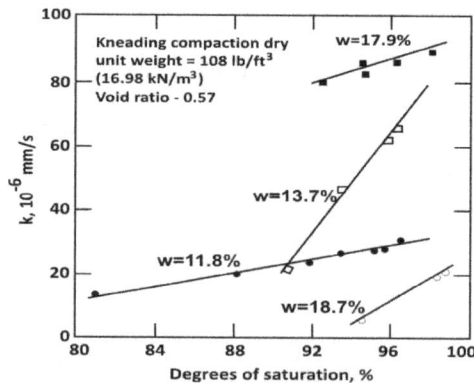

Influence of degree of saturation on permeability of compacted silty clay.
(Note: samples aged 21 days at constant water content and unit weight after compaction prior to test.) (Redrawn after J. K. Mitchell, D R. Hooper, and R. G. Campanella, Permeability of Compacted Clay. J. Soil Mech. Found. Div. ASCE, vol. 91, no. SM4 1965)

# Porosity

Porosity or void fraction is a measure of the void (i.e. "empty") spaces in a material, and is a fraction of the volume of voids over the total volume, between 0 and 1, or as a percentage between 0 and 100%. Strictly speaking, some tests measure the "accessible void", the total amount of void space accessible from the surface (cf. closed-cell foam). There are many ways to test porosity in a substance or part, such as industrial CT scanning. The term porosity is used in multiple fields including pharmaceutics, ceramics, metallurgy, materials, manufacturing, earth sciences, soil mechanics and engineering.

## Void Fraction in Two-phase Flow

In gas-liquid two-phase flow, the void fraction is defined as the fraction of the flow-channel volume that is occupied by the gas phase or, alternatively, as the fraction of the cross-sectional area of the channel that is occupied by the gas phase. Void fraction usually varies from location to location in the flow channel (depending on the two-phase flow pattern). It fluctuates with time and its value is usually time averaged. In separated (i.e., non-homogeneous) flow, it is related to volumetric flow rates of the gas and the liquid phase, and to the ratio of the velocity of the two phases (called *slip ratio*).

## Porosity in Earth Sciences and Construction

Used in geology, hydrogeology, soil science, and building science, the porosity of a porous medium (such as rock or sediment) describes the fraction of void space in the material, where the void may contain, for example, air or water. It is defined by the ratio:

$$\phi = \frac{V_V}{V_T}$$

where $V_V$ is the volume of void-space (such as fluids) and $V_T$ is the total or bulk volume of material, including the solid and void components. Both the mathematical symbols $\phi$ and $n$ are used to denote porosity.

Porosity is a fraction between 0 and 1, typically ranging from less than 0.01 for solid granite to more than 0.5 for peat and clay. It may also be represented in percent terms by multiplying the fraction by 100.

The porosity of a rock, or sedimentary layer, is an important consideration when attempting to evaluate the potential volume of water or hydrocarbons it may contain. Sedimentary porosity is a complicated function of many factors, including but not limited to: rate of burial, depth of burial, the nature of the connate fluids, the nature of overlying sediments (which may impede fluid expulsion). One commonly used relationship between porosity and depth is given by the Athy (1930) equation:

$$\phi(z) = \phi_0 e^{-kz}$$

where $\phi_0$ is the surface porosity, $k$ is the compaction coefficient (m⁻¹) and $z$ is depth (m).

A value for porosity can alternatively be calculated from the bulk density $\rho_{bulk}$, saturating fluid density $\rho_{fluid}$ and particle density $\rho_{particle}$:

$$\phi = \frac{\rho_{particle} - \rho_{bulk}}{\rho_{particle} - \rho_{fluid}}$$

If the void space if filled with air, the following simpler form may be used:

$$\phi = 1 - \frac{\rho_{bulk}}{\rho_{particle}}$$

Normal particle density is assumed to be approximately 2.65 g/cm³ (silica), although a better estimation can be obtained by examining the lithology of the particles.

## Porosity and Hydraulic Conductivity

Porosity can be proportional to hydraulic conductivity; for two similar sandy aquifers, the one with a higher porosity will typically have a higher hydraulic conductivity (more open area for the flow of water), but there are many complications to this relationship. The principal complication is that there is not a direct proportionality between porosity and hydraulic conductivity but rather an inferred proportionality. There is a clear proportionality between pore throat radii and hydraulic conductivity. Also, there tends to be a proportionality between pore throat radii and pore volume. If the proportionality between pore throat radii and porosity exists then a proportionality between porosity and hydraulic conductivity may exist. However, as grain size or sorting decreases the proportionality between pore throat radii and porosity begins to fail and therefore so does the proportionality between porosity and hydraulic conductivity. For example: clays typically have very low hydraulic conductivity (due to their small pore throat radii) but also have very high porosities (due to the structured nature of clay minerals), which means clays can hold a large volume of water per volume of bulk material, but they do not release water rapidly and therefore have low hydraulic conductivity.

## Sorting and Porosity

Well sorted (grains of approximately all one size) materials have higher porosity than similarly sized poorly sorted materials (where smaller particles fill the gaps between larger particles). The graphic illustrates how some smaller grains can effectively fill the pores (where all water flow takes place), drastically reducing porosity and hydraulic conductivity, while only being a small fraction of the total volume of the material.

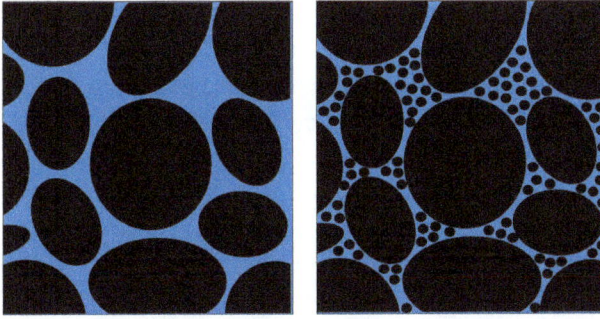

Effects of sorting on alluvial porosity. Black represents solids, blue represents pore space.

## Porosity of Rocks

Consolidated rocks (e.g., sandstone, shale, granite or limestone) potentially have more complex "dual" porosities, as compared with alluvial sediment. This can be split into connected and unconnected porosity. Connected porosity is more easily measured through the volume of gas or liquid that can flow into the rock, whereas fluids cannot access unconnected pores.

Porosity is the ratio of pore volume to its total volume. Porosity is controlled by: rock type, pore distribution, cementation, diagenetic history and composition. Porosity is not controlled by grain size, as the volume of between-grain space is related only to the method of grain packing.

Rocks normally decrease in porosity with age and depth of burial. Tertiary age Gulf Coast sandstones are in general more porous than Cambrian age sandstones. There are exceptions to this rule, usually because of the depth of burial and thermal history.

## Porosity of Soil

Porosity of surface soil typically decreases as particle size increases. This is due to soil aggregate formation in finer textured surface soils when subject to soil biological processes. Aggregation involves particulate adhesion and higher resistance to compaction. Typical bulk density of sandy soil is between 1.5 and 1.7 g/cm$^3$. This calculates to a porosity between 0.43 and 0.36. Typical bulk density of clay soil is between 1.1 and 1.3 g/cm$^3$. This calculates to a porosity between 0.58 and 0.51. This seems counterintuitive because clay soils are termed *heavy*, implying *lower* porosity. Heavy apparently refers to a gravitational moisture content effect in combination with terminology that harkens back to the relative force required to pull a tillage implement through the clayey soil at field moisture content as compared to sand.

Porosity of subsurface soil is lower than in surface soil due to compaction by gravity. Porosity of 0.20 is considered normal for unsorted gravel size material at depths below the biomantle. Porosity in finer material below the aggregating influence of pedogenesis can be expected to approximate this value.

Soil porosity is complex. Traditional models regard porosity as continuous. This fails to account for anomalous features and produces only approximate results. Furthermore, it cannot help model the influence of environmental factors which affect pore geometry. A number of more complex models have been proposed, including fractals, bubble theory, cracking theory, Boolean grain process, packed sphere, and numerous other models. The characterisation of pore space in soil is an associated concept.

## Types of Geologic Porosities

Primary porosity

> The main or original porosity system in a rock or unconfined alluvial deposit.

Secondary porosity

> A subsequent or separate porosity system in a rock, often enhancing overall porosity of a rock. This can be a result of chemical leaching of minerals or the generation of a fracture system. This can replace the primary porosity or coexist with it.

Fracture porosity

> This is porosity associated with a fracture system or faulting. This can create secondary porosity in rocks that otherwise would not be reservoirs for hydrocarbons due to their primary porosity being destroyed (for example due to depth of burial) or of a rock type not normally considered a reservoir (for example igneous intrusions or metasediments).

Vuggy porosity

> This is secondary porosity generated by dissolution of large features (such as macrofossils) in carbonate rocks leaving large holes, vugs, or even caves.

Effective porosity (also called *open porosity*)

> Refers to the fraction of the total volume in which fluid flow is effectively taking place and includes catenary and dead-end (as these pores cannot be flushed, but they can cause fluid movement by release of pressure like gas expansion) pores and excludes closed pores (or non-connected cavities). This is very important for groundwater and petroleum flow, as well as for solute transport.

Ineffective porosity (also called *closed porosity*)

> Refers to the fraction of the total volume in which fluids or gases are present but in which fluid flow can not effectively take place and includes the closed pores. Understanding the morphology of the porosity is thus very important for groundwater and petroleum flow.

Dual porosity

> Refers to the conceptual idea that there are two overlapping reservoirs which interact. In fractured rock aquifers, the rock mass and fractures are often simulated as being two overlapping but distinct bodies. Delayed yield, and leaky aquifer flow solutions are both mathematically similar solutions to that obtained for dual porosity; in all three cases water comes from two mathematically different reservoirs (whether or not they are physically different).

Macroporosity

> In solids (i.e. excluding aggregated materials such as soils), the term 'macroporosity' refers to pores greater than 50 nm in diameter. Flow through macropores is described by bulk diffusion.

Mesoporosity

> In solids (i.e. excluding aggregated materials such as soils), the term 'mesoporosity' refers to pores greater than 2 nm and less than 50 nm in diameter. Flow through mesopores is described by Knudsen diffusion.

Microporosity

> In solids (i.e. excluding aggregated materials such as soils), the term 'microporosity' refers to pores smaller than 2 nm in diameter. Movement in micropores is activated by diffusion.

## Porosity of Fabric or Aerodynamic Porosity

The ratio of holes to solid that the wind "sees". Aerodynamic porosity is less than visual porosity, by an amount that depends on the constriction of holes.

## Measuring Porosity

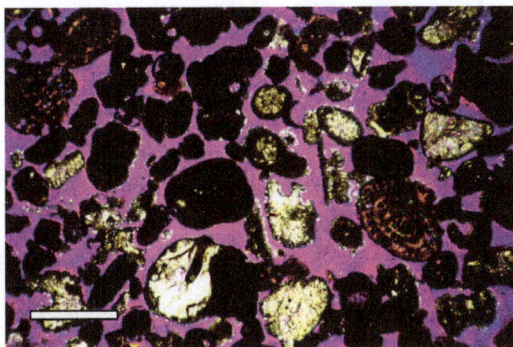

Optical method of measuring porosity: thin section under gypsum plate shows
porosity as purple color, contrasted with carbonate grains of other colors.
Pleistocene eolianite from San Salvador Island, Bahamas. Scale bar 500 μm.

Several methods can be employed to measure porosity:

- Direct methods (determining the bulk volume of the porous sample, and then determining the volume of the skeletal material with no pores (pore volume = total volume – material volume).

- Optical methods (e.g., determining the area of the material versus the area of the pores visible under the microscope). The "areal" and "volumetric" porosities are equal for porous media with random structure.

- Computed tomography method (using industrial CT scanning to create a 3D rendering of external and internal geometry, including voids. Then implementing a defect analysis utilizing computer software)

- Imbibition methods, i.e., immersion of the porous sample, under vacuum, in a fluid that preferentially wets the pores.

  o Water saturation method (pore volume = total volume of water – volume of water left after soaking).

- Water evaporation method (pore volume = (weight of saturated sample – weight of dried sample)/density of water)

- Mercury intrusion porosimetry (several non-mercury intrusion techniques have been developed due to toxicological concerns, and the fact that mercury tends to form amalgams with several metals and alloys).

- Gas expansion method. A sample of known bulk volume is enclosed in a container of known volume. It is connected to another container with a known volume which is evacuated (i.e., near vacuum pressure). When a valve connecting the two containers is opened, gas passes from the first container to the second until a uniform pressure distribution is attained. Using ideal gas law, the volume of the pores is calculated as

$$V_V = V_T - V_a - V_b \frac{P_2}{P_2 - P_1},$$

where

$V_V$ is the effective volume of the pores,

$V_T$ is the bulk volume of the sample,

$V_a$ is the volume of the container containing the sample,

$V_b$ is the volume of the evacuated container,

$P_1$ is the initial pressure in the initial pressure in volume $V_a$ and $V_V$, and

$P_2$ is final pressure present in the entire system.

The porosity follows straightforwardly by its proper definition

$$\phi = \frac{V_V}{V_T}.$$

Note that this method assumes that gas communicates between the pores and the surrounding volume. In practice, this means that the pores must not be closed cavities.

- Thermoporosimetry and cryoporometry. A small crystal of a liquid melts at a lower temperature than the bulk liquid, as given by the Gibbs-Thomson equation. Thus if a liquid is imbibed into a porous material, and frozen, the melting temperature will provide information on the pore-size distribution. The detection of the melting can be done by sensing the transient heat flows during phase-changes using differential scanning calorimetry – (DSC thermoporometry), measuring the quantity of mobile liquid using nuclear magnetic resonance – (NMR cryoporometry) or measuring the amplitude of neutron scattering from the imbibed crystalline or liquid phases – (ND cryoporometry).

## Void Ratio

Void ratio, in materials science, is a quantity related to porosity and defined as the ratio:

$$e = \frac{V_V}{V_S} = \frac{V_V}{V_T - V_V} = \frac{\phi}{1-\phi}$$

and

$$\phi = \frac{V_V}{V_T} = \frac{V_V}{V_S + V_V} = \frac{e}{1+e}$$

where $e$ is void ratio, $\phi$ is porosity, $V_V$ is the volume of void-space (such as fluids), $V_S$ is the volume of solids, and $V_T$ is the total or bulk volume. This figure is relevant in composites, in mining (particular with regard to the properties of tailings), and in soil science. In geotechnical engineering, it is considered as one of the state variables of soils and represented by the symbol $e$.

Note that in geotechnical engineering, the symbol $\phi$ usually represents the angle of shearing resistance, a shear strength (soil) parameter. Because of this, the equation is usually rewritten using $n$ for porosity:

$$e = \frac{V_V}{V_S} = \frac{V_V}{V_T - V_V} = \frac{n}{1-n}$$

and

$$n = \frac{V_V}{V_T} = \frac{V_V}{V_S + V_V} = \frac{e}{1+e}$$

where $e$ is void ratio, $n$ is porosity, $V_v$ is the volume of void-space (air and water), $V_s$ is the volume of solids, and $V_T$ is the total or bulk volume.

## Engineering Applications

- Volume change tendency control. If void ratio is high (loose soils) voids in a soil skeleton tend to minimize under loading - adjacent particles contract. The opposite situation, i.e. when void ratio is relatively small (dense soils), indicates that the volume of the soil is vulnerable to increase under loading - particles dilate.

- Fluid conductivity control (ability of water movement through the soil). Loose soils show high conductivity, while dense soils are not so permeable.

- Particles movement. In a loose soil particles can move quite easily, whereas in a dense one finer particles cannot pass through the voids, which leads to clogging.

## Permeability for Stratified Soils

In general, natural soil deposits are stratified. If the stratification is continuous, the effective coefficients of permeability for flow in the horizontal and vertical directions can be readily calculated.

Flow in the horizontal direction. Figure shows several layers of soil with horizontal stratification. Due to fabric anisotropy, the coefficient of permeability of each soil layer may vary depending on the direction of flow. So, let us assume that $k_{h_1}, k_{h_2}, k_{h_3}, \ldots \ldots$ are the coefficients of permeability of layers 1, 2, 3, ...., respectively, for flow in the horizontal direction. Similarly, let $k_{v_1}, k_{v_2}, k_{v_3}, \ldots \ldots$ be the coefficients of permeability for flow in the vertical direction.

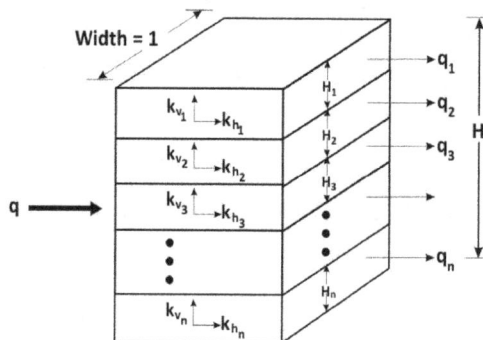

Flows in horizontal direction is stratified soil deposit

Considering unit width of the soil layers as shown in Figure, the rate of seepage in the horizontal direction can be given by

$$q = q_{\text{Đ}} + q\ + q\ + \ddot{u} + q_n \tag{13}$$

Where q is the flow rate through the stratified soil layers combined, and $q_1 + q_2 + q_3 \dots$ is the rate of flow through soil layers 1, 2, 3,...., Respectively. Note that for flow in the horizontal direction (which is the direction of stratification of the soil layers).the hydraulic gradient is the same for all layers. So,

$$\begin{aligned} q_1 &= k_{h_1}, iH_1 \\ q_2 &= k_{h_2}, iH_2 \\ q_3 &= k_{h_3}, iH_3 \end{aligned} \tag{14}$$

$$\vdots$$

And $q = k_{e(h)} iH$ \hfill (15)

Where

$i = hydraulic\ gradient$

$k_{e(h)} = effective\ coefficient\ of\ permeability\ for\ flow\ in\ horizontal\ direction$

$H = H_1 + H_2 + H_3 \dots \dots$

Substitution of equation (14) and (15) into equation (13) yields

$$k_{e(h)} H = k_{h_1} H_1 + k_{h_2} H_2 + k_{h_3} H_3 \dots$$

$$Hence, k_{e(h)} = \frac{1}{H}(k_{h_1} H_1 + k_{h_2} H_2 + k_{h_3} H_3 \dots) \tag{16}$$

Flow in the vertical direction. For flow in the vertical direction for the soil layers shown in Figure.

$$v = v_1 = v_2 = v_3 = \cdots = v_n \tag{17}$$

Where $v_1, v_2, v_2 \dots$ .. are the discharge velocities in layers 1, 2, 3, ...., respectively; or

$$v = k_{e(v)} i = k_{v(1)} i_1 = k_{v(2)} i_2 = k_{v(3)} i_3 = \cdots \tag{18}$$

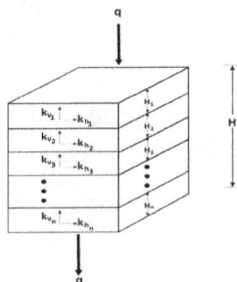

Flow in vertical direction in stratified soil deposit

Where

$k_{e(v)}$ = effective coefficient of permeability for flow in vertical direction

$k_{v(1)}k_{v(2)}k_{v(3)}......=$
coefficients of permeability of layers 1,2,3, respectively for flow in vertical direction
$i_1,i_2,i_3....=$ hydraulic gradient in soil layers 1,2,3,......,respectively

For flow at right angles to the direction of stratification,

Total head = (head loss in layer 1 + (head loss in layer 2) +........or

$$iH = i_1 H_1 + i_2 H_2 + i_3 H_3 + \cdots$$

(19)

Combining equation (18) and (19)

$$\frac{v}{k_{e(v)}}H = \frac{v}{k_{v1}}H_1 + \frac{v}{k_{v2}}H_2 + \frac{v}{k_{v3}}H_3 + or...$$

$$k_{e(v)} = \frac{H}{H_1/k_{v1} + H_2/k_{v2} + H_3/k_{v3} + \cdots}$$

(20)

## Determination of Coefficient in the Laboratory

The four most common laboratory methods for determining the coefficient of permeability of soils are the following:

1. Constant-head test.

2. Falling-head test.

3. Indirect determination form consolidation test.

4. Indirect determination by horizontal capillary test.

The general principles of these methods are given below.

1. Constant-head test : The constant-head test is suitable for more permeable granular materials. The basic laboratory test arrangement is shown in Figure. The soil specimen is placed inside a cylindrical mold, and the constant head loss, h, of water flowing through the soil is maintained by adjusting the supply. The outflow water is collected in a measuring cylinder, and the duration of the collection period is noted. From Darcy's law, the total quantity of flow Q in time t can be given by

$$Q = qt = kiAt$$

Where A is the area of cross section of the specimen. But $i = h / L$, where L is the length of the specimen, and so $Q = k(h / L)At$. Rearranging this gives

$$k = \frac{QL}{hAt} \qquad (21)$$

Once all the quantities in the right-hand side of equation (21) have been determined from the test, the coefficient of permeability of the soil can be calculated.

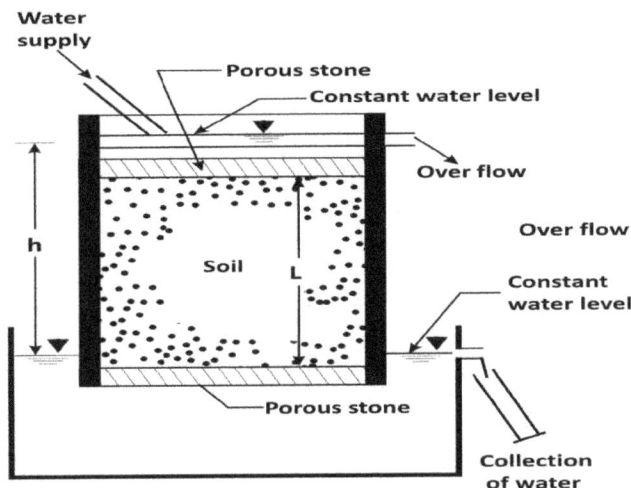

Constant-head laboratory permeability test

2. Falling-head test : The falling-head permeability test is more suitable for fine-grained soils. Figure shows the general laboratory arrangement for the test. The soil specimen is placed inside a tube, and a standpipe is attached to the top of the specimen. Water from the standpipe flows through the specimen. The initial head difference $h_1$ at time $t = 0$ is recorded, and water is allowed to flow through the soil such that the final head difference at time $t = t$ is $h_2$.

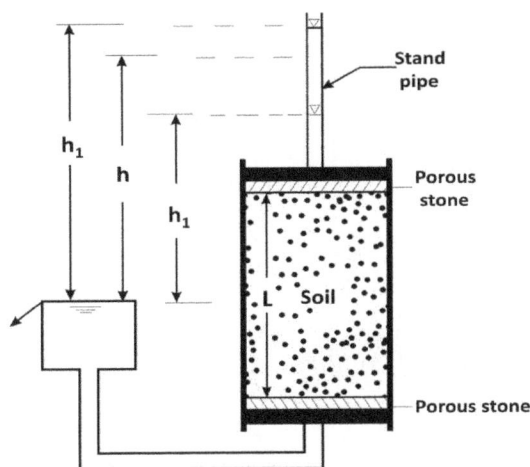

Falling-head laboratory permeability test

$$q = kiA = k\frac{h}{L}A = -a\frac{dh}{dt}$$

(22)

Where

> $h$ = head difference at any time $t$
> $A$ = area of specimen
> $a$ = area of standpipe
> $L$ = length of specimen

From equation (22)

$$\int_0^t dt = \int_{h_1}^{h_2} \frac{aL}{Ak}\left(-\frac{dh}{h}\right)$$

Or $k = 2.303\frac{aL}{At}\log\frac{h_1}{h_2}$

(23)

The values of $a, L, A, t, h_1,$ and $h_2$ can be determined from the test, and then the coefficient of the permeability k for a soil can be calculated from equation (23).

3. Permeability form consolidation test : The coefficient of permeability of clay soils is often determined by the consolidation test.

$$T_v = \frac{C_v t}{H^2}$$

Where

> $T_v$ = time factor
> $C_v$ = coefficient of consolidation
> $H$ = lenght of average drainage path
> $t$ = time

The coefficient of consolidation is

$$C_v = \frac{k}{\gamma_w m_v}$$

Where

> $\gamma_w$ = unit weight of water
> $m_v$ = volume coefficient of compressibility

Also, $m_v = \dfrac{\Delta e}{\Delta\sigma(1+e)}$

Where

$\Delta e = change\ of\ void\ ratio\ for\ incremental\ loading$

$\Delta\sigma = incremental\ pressure\ applied$

$e = initial\ void\ ratio$

Combining these three equations, we have

$$k = \frac{T_v \gamma_w \Delta e H^2}{t\Delta\sigma(1+e)} \tag{24}$$

For 50% consolidation, $T_v = 0.198$ ; and the corresponding $t_{50}$ can be estimated according to the procedure presented.

$$Hence, k = \frac{0.198\gamma_w \Delta e\, H^2}{t_{50}\Delta\sigma(1+e)} \tag{25}$$

4. Horizontal capillary test : The fundamental principle behind the horizontal capillary test can be explained with the aid of Figure, which shows an initially dry soil inside a horizontal tube. If the valve A is opened, water from the reservoir will enter the tube and, through capillary action, the line of the wetted surface in the soil will gradually advance-in order words, the distance x from the point 1 is a function of time t.

At point 1, the total head is zero (based on the datum shown in Figure). At point 2 (immediately to the left of the wetted surface), the total head is $- (h+h_c)$. Using Darcy's law,

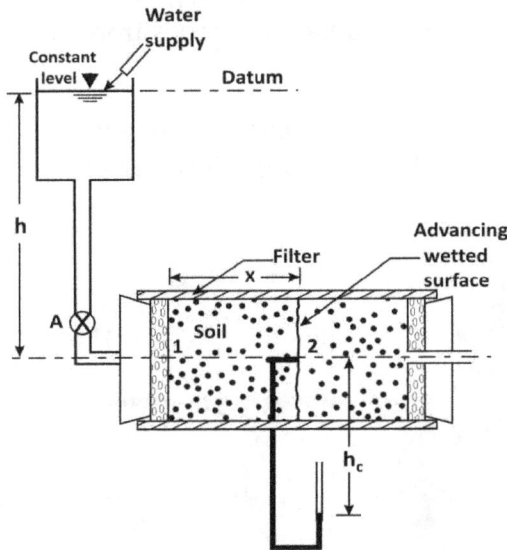

Development of horizontal capillary test

$$v = nS_r v_s = ki \tag{26}$$

Where

$n = porosity$

$S_r = degree\,of\,saturation$

$v_s = seepage\,of\,velocity$

But $v_s = \dfrac{dx}{dt}$

And $i = \dfrac{(total\,head\,at\,1)-(total\,head\,at\,2)}{x}$ \qquad (27)

Or $i = \dfrac{0-[-(h+h_c)]}{x} = \dfrac{h+h_c}{x}$ \qquad (28)

Substituting equation (27) and (28) into (26), we get

$$v_s = \frac{dx}{dt} = k\frac{1}{nS_r}\frac{h+h_c}{x}$$

$$\int_{x_1}^{x_2} x\,dx = \int_0^t \frac{k}{nS_r}(h+h_c)dt$$

$$\frac{x_2^2-x_1^2}{t} = \frac{2k}{nS_r}(h+h_c)$$ \qquad (29)

Equation (29) is the basic relation used for determination of the coefficient, of permeability. The degree of saturation of the soil during the movement of the water front is sometimes assumed to be 100%. In fact, $S_r$ varies from about 75 to 95% for tests in most soils.

Figure shows the general laboratory arrangement for a horizontal capillary test. A brief outline of the steps for conducting the test is given below.

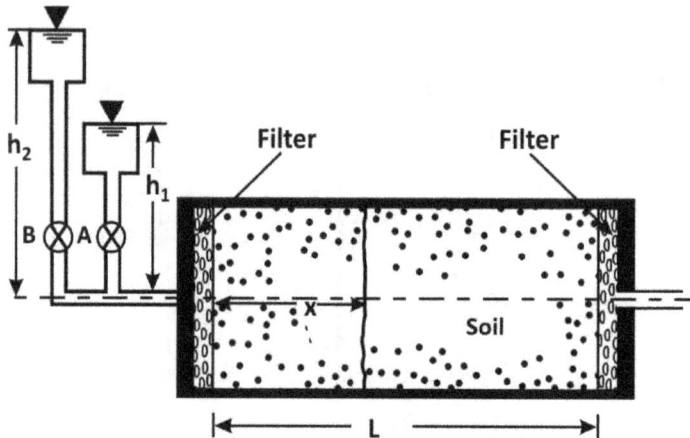

Horizontal capillary permeability test

Open the valve A.

As the water front gradually travels forward, note the elapsed times t and the corresponding distances x traveled by the water front.

When the water front has traveled about half the length of the sample (i.e., when x is about L/2), close valve A and open valve B.

Continue to note the advance of water front with time, until x is equal to L.

Close valve B. removes the soil specimen and determine the moisture content and the degree of saturation.

Plot the values of $x^2$ against the corresponding time t. Figure shows the nature of the plot, which consists of two straight lines. The portion $0a$ is for the readings taken in step 2, and the portion $ab$ is for the readings taken in step 4.

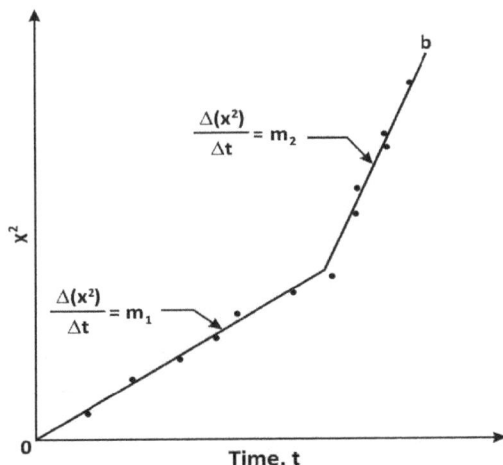

Plot x² against time t in horizontal capillary permeability test

From equation (29) we can write

$$\frac{\Delta(x^2)}{\Delta t} = \frac{2k}{nS_r}(h+h_c) \tag{30}$$

The left-hand side of equation (30) represents the slope of the straight-line plot of x² vs t.

Determine the slopes of lines $0a$ and $ab$. Let these be $m_1$ and $m_2$, respectively. So,

$$m_1 = \frac{2k}{nS_r}(h_1 + h_c) \tag{31}$$

$$and\, m_2 = \frac{2k}{nS_r}(h_2 + h_c) \tag{32}$$

since $n, S_r, h_1, h_2, m_1 \, and \, m_2$ are determined from the test, the above two equation contain only two unknowns ( $k \, and \, h_c$ ) and thus can be solved.

## References

- Glasbey, C. A.; G. W. Horgan; J. F. Darbyshire (September 1991). "Image Analysis And Three-Dimensional Modelling Of Pores In Soil Aggregates". Journal Of Soil Science. 42 (3): 479–86. Doi:10.1111/J.1365-2389.1991.Tb00424.X

- Iso (1995). "Soil Quality -- Determination Of Pore Water Pressure -- Tensiometer Method". Iso 11276:1995. International Standards Organization. Retrieved 2014-03-13

- Ringrose-Voase, A.j.; Sanidad, W.b. (1996). "A Method For Measuring The Development Of Surface Cracks In Soils: Application To Crack Development After Lowland Rice". Geoderma. 71: 245–261. Doi:10.1016/0016-7061(96)00008-0

- Wang, H. F., 2000. Theory Of Linear Poroelasticity With Applications To Geomechanics And Hydrogeology, Princeton University Press. Isbn 0-691-03746-9

- Horgan, G. W. (June 1998). "Mathematical Morphology For Soil Image Analysis". European Journal Of Soil Science. 49 (2): 161–73. Doi:10.1046/J.1365-2389.1998.00160.X

- Brinkman, H. C. (1949). "A Calculation Of The Viscous Force Exerted By A Flowing Fluid On A Dense Swarm Of Particles". Applied Scientific Research. 1: 27–34. Doi:10.1007/Bf02120313

- Soil Science Glossary Terms Committee (2008). Glossary Of Soil Science Terms 2008. Madison, Wi: Soil Science Society Of America. Isbn 978-0-89118-851-3

- Materials Engineering And Research Laboratory. "Procedure For Using Piezometers To Monitor Water Pressure In A Rock Mass" (Pdf). Usbr 6515. U.s. Bureau Of Reclamation. Retrieved 2014-03-13

- Kerkhof, P. (1996). "A Modified Maxwell–Stefan Model For Transport Through Inert Membranes: The Binary Friction Model". Chemical Engineering Journal And The Biochemical Engineering Journal. 64: 319–343. Doi:10.1016/S0923-0467(96)03134-X

- Jin, Y.; Uth, M.-F.; Kuznetsov, A. V.; Herwig, H. (2 February 2015). "Numerical Investigation Of The Possibility Of Macroscopic Turbulence In Porous Media: A Direct Numerical Simulation Study". Journal Of Fluid Mechanics. 766: 76–103. Bibcode:2015Jfm...766...76J. Doi:10.1017/Jfm.2015.9

- Coduto, Donald; Et Al. (2011). Geotechnical Engineering Principles And Practices. Nj: Pearson Higher Education, Inc. Isbn 9780132368681

# Shear Strength of Soil

Shear strength is the stress felt on the cross-section of a surface. Shear strength of soil analyzes the shear stress that soil can maintain. This chapter serves as a source to understand the major categories related to soil texture. Geotechnical engineering is best understood in confluence with the major topics listed in the following chapter.

## Shear Strength

The shear strength of soils is an important aspect in many geotechnical engineering engineering problems such as the bearing capacity of foundations, the stability of the slopes of dams and embankments, and lateral earth pressure on retaining walls.

In engineering, shear strength is the strength of a material or component against the type of yield or structural failure where the material or component fails in shear. A shear load is a force that tends to produce a sliding failure on a material along a plane that is parallel to the direction of the force. When a paper is cut with scissors, the paper fails in shear.

In structural and mechanical engineering, the shear strength of a component is important for designing the dimensions and materials to be used for the manufacture or construction of the component (e.g. beams, plates, or bolts). In a reinforced concrete beam, the main purpose of reinforcing bar (rebar) stirrups is to increase the shear strength.

For shear stress $\tau$ applies

$$\tau = \frac{\sigma_1 - \sigma_3}{2},$$

where

$\sigma_1$ is major principal stress and

$\sigma_3$ is minor principal stress.

In general: ductile materials (e.g. aluminium) fail in shear, whereas brittle materials (e.g. cast iron) fail in tension.

To calculate:

Given total force at failure (F) and the force-resisting area (e.g. the cross-section of a bolt loaded in shear), ultimate shear strength ($\tau$) is:

$$\tau = \frac{F}{A} = \frac{F}{\pi r_{bolt}^2} = \frac{4F}{\pi d_{bolt}^2}$$

## Comparison

As a very rough guide relating tensile, yield, and shear strengths:

| Material | Ultimate Strength Relationship | Yield Strength Relationship |
|---|---|---|
| Steels | USS = approx. 0.75*UTS | SYS = approx. 0.58*TYS |
| Ductile Iron | USS = approx. 0.9*UTS | SYS = approx. 0.75*TYS |
| Malleable Iron | USS = approx. 1.0*UTS | |
| Wrought Iron | USS = approx. 0.83*UTS | |
| Cast Iron | USS = approx. 1.3*UTS | |
| Aluminums | USS = approx. 0.65*UTS | SYS = approx. 0.55*TYS |

USS: Ultimate Shear Strength, UTS: Ultimate Tensile Strength, SYS: Shear Yield Stress, TYS: Tensile Yield Stress

| Material | Ultimate stress (Ksi) | Ultimate stress (MPa) |
|---|---|---|
| Fiberglass/epoxy (23 ° C) | 7.82 | 53.9 |

When values measured from physical samples are desired, a number of testing standards are available, covering different material categories and testing conditions. In the US, ASTM standards for measuring shear strength include ASTM B831, D732, D4255, D5379, and D7078. Internationally, ISO testing standards for shear strength include ISO 3597, 12579, and 14130.

## Shear Strength (Soil)

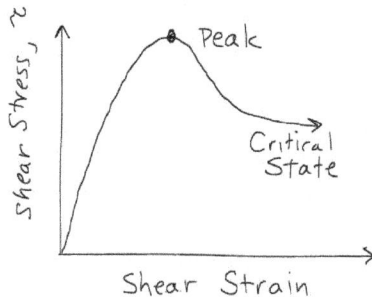

Typical stress strain curve for a drained dilatant soil

Shear strength is a term used in soil mechanics to describe the magnitude of the shear stress that a soil can sustain. The shear resistance of soil is a result of friction and interlocking of particles, and possibly cementation or bonding at particle contacts. Due to interlocking, particulate material may expand or contract in volume as it is subject to shear strains. If soil expands its volume, the density of particles will decrease and the strength will decrease; in this case, the peak strength would be followed by a reduction of shear stress. The stress-strain relationship levels off when the material stops expanding or contracting, and when interparticle bonds are broken. The theoretical state at which the shear stress and density remain constant while the shear strain increases may be called the critical state, steady state, or residual strength.

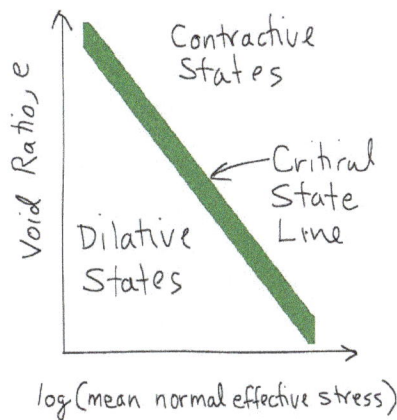

A critical state line separates the dilatant and contractive states for soil

The volume change behavior and interparticle friction depend on the density of the particles, the intergranular contact forces, and to a somewhat lesser extent, other factors such as the rate of shearing and the direction of the shear stress. The average normal intergranular contact force per unit area is called the effective stress.

If water is not allowed to flow in or out of the soil, the stress path is called an *undrained stress path*. During undrained shear, if the particles are surrounded by a nearly incompressible fluid such as water, then the density of the particles cannot change without drainage, but the water pressure and effective stress will change. On the other hand, if the fluids are allowed to freely drain out of the pores, then the pore pressures will remain constant and the test path is called a *drained stress path*. The soil is free to dilate or contract during shear if the soil is drained. In reality, soil is partially drained, somewhere between the perfectly undrained and drained idealized conditions.

The shear strength of soil depends on the effective stress, the drainage conditions, the density of the particles, the rate of strain, and the direction of the strain.

For undrained, constant volume shearing, the Tresca theory may be used to predict the shear strength, but for drained conditions, the Mohr–Coulomb theory may be used.

Two important theories of soil shear are the critical state theory and the steady state theory. There are key differences between the critical state condition and the steady state condition and the resulting theory corresponding to each of these conditions.

## Factors Controlling Shear Strength of Soils

The stress-strain relationship of soils, and therefore the shearing strength, is affected (Poulos 1989) by:

1. soil composition (basic soil material): mineralogy, grain size and grain size distribution, shape of particles, pore fluid type and content, ions on grain and in pore fluid.

2. state (initial): Defined by the initial void ratio, effective normal stress and shear stress (stress history). State can be described by terms such as: loose, dense, overconsolidated, normally consolidated, stiff, soft, contractive, dilative, etc.

3. structure: Refers to the arrangement of particles within the soil mass; the manner the particles are packed or distributed. Features such as layers, joints, fissures, slickensides, voids, pockets, cementation, etc., are part of the structure. Structure of soils is described by terms such as: undisturbed, disturbed, remolded, compacted, cemented; flocculent, honey-combed, single-grained; flocculated, deflocculated; stratified, layered, laminated; isotropic and anisotropic.

4. Loading conditions: Effective stress path, i.e., drained, and undrained; and type of loading, i.e., magnitude, rate (static, dynamic), and time history (monotonic, cyclic).

## Undrained Strength

This term describes a type of shear strength in soil mechanics as distinct from drained strength.

Conceptually, there is no such thing as *the* undrained strength of a soil. It depends on a number of factors, the main ones being:

- Orientation of stresses

- Stress path

- Rate of shearing

- Volume of material (like for fissured clays or rock mass)

Undrained strength is typically defined by Tresca theory, based on Mohr's circle as:

$$\sigma_1 - \sigma_3 = 2 S_u$$

Where:

$\sigma_1$ is the major principal stress

$\sigma_3$ is the minor principal stress

$\tau$ is the shear strength $(\sigma_1 - \sigma_3)/2$

hence, $\tau = S_u$ (or sometimes $c_u$), the undrained strength.

It is commonly adopted in limit equilibrium analyses where the rate of loading is very much greater than the rate at which pore water pressures, that are generated due to the action of shearing the soil, may dissipate. An example of this is rapid loading of sands during an earthquake, or the failure of a clay slope during heavy rain, and applies to most failures that occur during construction.

As an implication of undrained condition, no elastic volumetric strains occur, and thus Poisson's ratio is assumed to remain 0.5 throughout shearing. The Tresca soil model also assumes no plastic volumetric strains occur. This is of significance in more advanced analyses such as in finite element analysis. In these advanced analysis methods, soil models other than Tresca may be used to model the undrained condition including Mohr-Coulomb and critical state soil models such as the modified Cam-clay model, provided Poisson's ratio is maintained at 0.5.

One relationship used extensively by practicing engineers is the empirical observation that the ratio of the undrained shear strength c to the original consolidation stress p' is approximately a constant for a given Over Consolidation Ratio (OCR). This relationship was first formalized by (Henkel 1960) and (Henkel & Wade 1966) who also extended it to show that stress-strain characteristics of remolded clays could also be normalized with respect to the original consolidation stress. The constant c/p relationship can also be derived from theory for both critical-state and steady-state soil mechanics (Joseph 2012). This fundamental, normalization property of the stress-strain curves is found in many clays, and was refined into the empirical SHANSEP (stress history and normalized soil engineering properties) method.(Ladd & Foott 1974).

## Drained Shear Strength

The drained shear strength is the shear strength of the soil when pore fluid pressures, generated during the course of shearing the soil, are able to dissipate during shearing. It also applies where no pore water exists in the soil (the soil is dry) and hence pore fluid pressures are negligible. It is commonly approximated using the Mohr-Coulomb equation. (It was called "Coulomb's equation" by Karl von Terzaghi in 1942.) (Terzaghi 1942) combined it with the principle of effective stress.

In terms of effective stresses, the shear strength is often approximated by:

$$\tau = \sigma' \, tan(\varphi') + c'$$

Where $\sigma' = (\sigma - u)$, is defined as the effective stress. $\sigma$ is the total stress applied normal to the shear plane, and $u$ is the pore water pressure acting on the same plane.

$\varphi'$ = the effective stress friction angle, or the 'angle of internal friction' after Coulomb friction. The coefficient of friction $\mu$ is equal to $\tan(\varphi')$. Different values of friction angle can be defined, including the peak friction angle, $\varphi'_p$, the critical state friction angle, $\varphi'_{cv}$, or residual friction angle, $\varphi'_r$.

$c'$ = is called cohesion, however, it usually arises as a consequence of forcing a straight line to fit through measured values of $(\tau, \sigma')$ even though the data actually falls on a curve. The intercept of the straight line on the shear stress axis is called the cohesion. It is well known that the resulting intercept depends on the range of stresses considered: it is not a fundamental soil property. The curvature (nonlinearity) of the failure envelope occurs because the dilatancy of closely packed soil particles depends on confining pressure.

## Critical State Theory

A more advanced understanding of the behaviour of soil undergoing shearing lead to the development of the critical state theory of soil mechanics (Roscoe, Schofield & Wroth 1958). In critical state soil mechanics, a distinct shear strength is identified where the soil undergoing shear does so at a constant volume, also called the 'critical state'. Thus there are three commonly identified shear strengths for a soil undergoing shear:

- Peak strength $\tau_p$
- Critical state or constant volume strength $\tau_{cv}$
- Residual strength $\tau_r$

The peak strength may occur before or at critical state, depending on the initial state of the soil particles being sheared:

- A loose soil will contract in volume on shearing, and may not develop any peak strength above critical state. In this case 'peak' strength will coincide with the critical state shear strength, once the soil has ceased contracting in volume. It may be stated that such soils do not exhibit a distinct 'peak strength'.

- A dense soil may contract slightly before granular interlock prevents further contraction (granular interlock is dependent on the shape of the grains and their initial packing arrangement). In order to continue shearing once granular interlock has occurred, the soil must dilate (expand in volume). As additional shear force is required to dilate the soil, a 'peak' strength occurs. Once this peak strength caused by dilation has been overcome through continued shearing, the resistance provided by the soil to the applied shear stress reduces (termed "strain softening"). Strain softening will continue until no further changes in volume of the soil occur on continued shearing. Peak strengths are also observed in overconsolidated clays where the natural fabric of the soil must be destroyed prior to reaching constant volume shearing. Other effects that result in peak strengths include cementation and bonding of particles.

The constant volume (or critical state) shear strength is said to be extrinsic to the soil, and independent of the initial density or packing arrangement of the soil grains. In this state the grains being sheared are said to be 'tumbling' over one another, with no significant granular interlock or sliding plane development affecting the resistance to shearing. At this point, no inherited fabric or bonding of the soil grains affects the soil strength.

The residual strength occurs for some soils where the shape of the particles that make up the soil become aligned during shearing (forming a slickenside), resulting in reduced resistance to continued shearing (further strain softening). This is particularly true for most clays that comprise plate-like minerals, but is also observed in some granular soils with more elongate shaped grains. Clays that do not have plate-like minerals (like allophanic clays) do not tend to exhibit residual strengths.

Use in practice: If one is to adopt critical state theory and take $c' = 0$; $\tau_p$ may be used, provided the level of anticipated strains are taken into account, and the effects of potential rupture or strain softening to critical state strengths are considered. For large strain deformation, the potential to form slickensided surface with a $\varphi'_r$ should be considered (such as pile driving).

The Critical State occurs at the quasi-static strain rate. It does not allow for differences in shear strength based on different strain rates. Also at the critical state, there is no particle alignment or specific soil structure.

Almost as soon as it was first introduced, the critical state concept has been subject to much criticism--chiefly its inability to match readily available test data from testing a wide variety of soils. This is primarily due to the theories inability to account for particle structure. A major consequence of this is its inability to model strain-softening post peak commonly observed in contractive soils that have anisotropic grain shapes/properties. Further, an assumption commonly made to make the model mathematically tractable is that shear stress cannot cause volumetric strain nor volumetric stress cause shear strain. Since this is not the case in reality, it is an additional cause of the poor matches to readily available empirical test data. Additionally, critical state elasto-plastic models assume that elastic strains drives volumetric changes. Since this too is not the case in real soils, this assumption results in poor fits to volume and pore pressure change data.

## Steady State (Dynamical Systems based Soil Shear)

A refinement of the critical state concept is the steady state concept.

The steady state strength is defined as the shear strength of the soil when it is at the steady state condition. The steady state condition is defined (Poulos 1981) as "that state in which the mass is continuously deforming at constant volume, constant normal effective stress, constant shear stress, and constant velocity." Steve J. Poulos, then an Associate Professor of the Soil Mechanics Department of Harvard University, built off a hypothesis that Arthur Casagrande was formulating towards the end of his career.(Poulos 1981) Steady state based soil mechanics is sometimes called "Harvard soil mechanics". The steady state condition is not the same as the "critical state" condition.

The steady state occurs only after all particle breakage if any is complete and all the particles are oriented in a statistically steady state condition and so that the shear stress needed to continue deformation at a constant velocity of deformation does not change. It applies to both the drained and the undrained case.

The steady state has a slightly different value depending on the strain rate at which it is measured. Thus the steady state shear strength at the quasi-static strain rate (the strain rate at which the critical state is defined to occur at) would seem to correspond to the critical state shear strength. However, there is an additional difference between the two states. This is that at the steady state condition the grains position themselves in the steady state structure, whereas no such structure occurs for the critical state. In the case of shearing to large strains for soils with elongated particles, this steady state structure is one where the grains are oriented (perhaps even aligned) in the direction of shear. In the case where the particles are strongly aligned in the direction of shear, the steady state corresponds to the "residual condition."

Three common misconceptions regarding the steady state are that a) it is the same as the critical state (it is not), b) that it applies only to the undrained case (it applies to all forms of drainage), and c) that it does not apply to sands (it applies to any granular material). A primer on the Steady State theory can be found in a report by Poulos (Poulos 1971). Its use in earthquake engineering is described in detail in another publication by Poulos (Poulos 1989).

The difference between the steady state and the critical state is not merely one of semantics as is sometimes thought, and it is incorrect to use the two terms/concepts interchangeably. The additional requirements of the strict definition of the steady state over and above the critical state viz. a constant deformation velocity and statistically constant structure (the steady state structure), places the steady state condition within the framework of dynamical systems theory. This strict definition of the steady state was used to describe soil shear as a dynamical system (Joseph 2012). Dynamical systems are ubiquitous in nature (the Great Red Spot on Jupiter is one example) and mathematicians have extensively studied such systems. The underlying basis of the soil shear dynamical system is simple friction (Joseph 2017).

# Shear Strength of Cohesive Soil

The shear strength of cohesive soils can, generally, be determined in the laboratory by either direct shear test equipment or triaxial shear test equipment; however, the triaxial test is more commonly used. Only the shear strength of saturated cohesive soils will be treated here. The shear strength based on the effective stress can be given by [equation 3] $s = c + \sigma' \tan \phi$. For normally consolidated clays, $c \approx 0$; and, for overconsolidated clays, $c > 0$.

## Triaxial Testing in Clays

The basic features of the triaxial test equipment were shown in Figure. Three conventional types of tests are conducted with clay soils in the laboratory:

1. Consolidated drained test or drained test (CD test or d test).

2. Consolidated undrained test (CU test).

3. Unconsolidated undrained test (UU test).

Each of these tests will be separately considered in the following sections.

> Consolidated drained test

For the consolidated drained test the saturated soil specimen is first subjected to a confining pressure $\sigma_3$ through the chamber fluid; as a result, the pore water pressure of the sample will increase by $u_c$. The connection to the drainage is kept open for complete drainage so that $u_c$ becomes equal to zero. Then the deviator stress (piston stress) $\Delta\sigma$ is increased at a very slow rate, keeping the drainage valve open to allow complete dissipation of the resulting pore water pressure $u_d$. Figure shows the nature of the variation of the deviator stress with axial strain. From Figure, it must also be pointed out that, during the application of the deviator stress, the volume of the specimen gradually reduces for normally consolidated clays. However, overconsolidated clays go through some reduction of volume initially but then expand. In a consolidated drained test, the total stress is equal to the effective stress since the excess pore water pressure is zero. At failure, the maximum effective principal stress is $\sigma'_1 = \sigma_1 = \sigma_3 + \Delta\sigma_f$, where $\Delta\sigma_f$ is the deviator stress at failure. The minimum effective principal stress is $\sigma'_3 = \sigma_3$.

From the results of a number of tests conducted using several specimens, the Mohr's circles at failure can be plotted as shown in Figure. The values of c and $\varnothing$ are obtained by drawing a common tangent to these Mohr's circles, which is the Mohr-Coulomb envelope. For normally consolidated clays, we can see that $c = 0$. Thus the equation of the Mohr-Coulomb envelope can be given by $s = \sigma' \tan \phi$. The slope of the failure envelope will give us the angle of friction of the soil. As shown by equation for these soils

$$\sin \phi = \left( \frac{\sigma'_1 - \sigma'_3}{\sigma'_1 + \sigma'_3} \right)_{failure} \quad or \quad \sigma'_1 = \sigma'_3 \tan^2 \left( 45° + \frac{\phi}{2} \right)$$

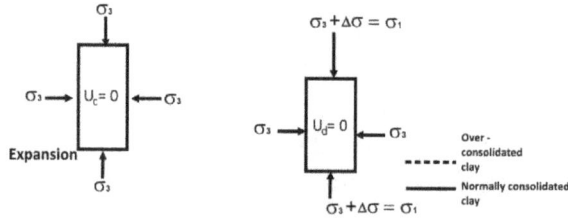

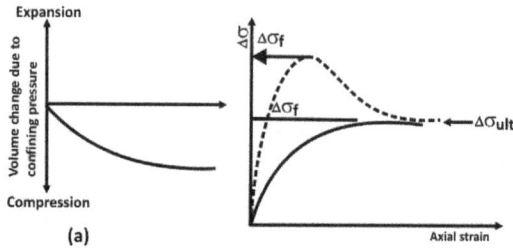

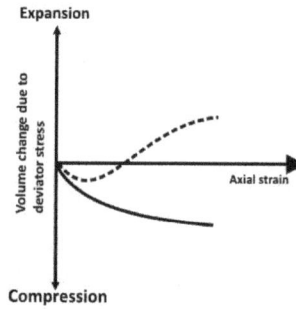

Consolidation drained triaxial tests in clay
(a) application of confining pressure (b) application of deviator stress

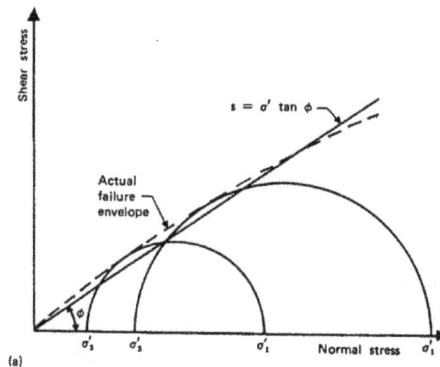

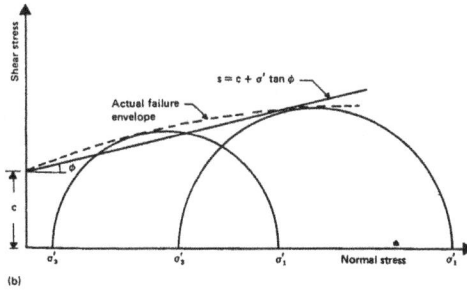

Failure envelopes for (a) normally consolidated and
(b) overconsolidated clays from consolidated drained triaxial tests

The plane of failure makes an angle of $45° + \phi/2$ with the major principal plane.

For overconsolidated clays $c \neq 0$. So, the shear strength follows the equation $s = c + \sigma'\tan\phi$ the values of $c$ and $\phi$ can be determined by measuring the intercept of the failure envelope on the shear stress axis and the slope of the failure envelope, respectively. To obtain a general relation between $\sigma'_1, \sigma'_3, c,$ and $\phi$, refer to Figure, from which

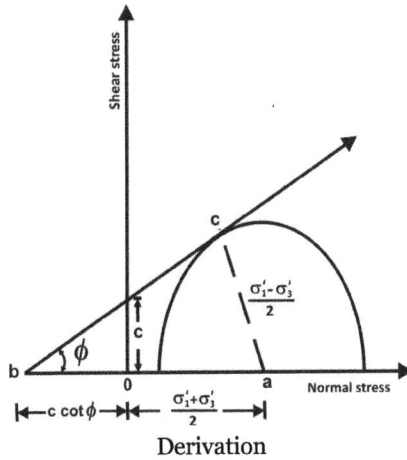

Derivation

$$\sin\phi = \frac{\overline{ac}}{\overline{bo}+\overline{Oa}} = \frac{(\sigma'_1-\sigma'_3)/2}{c\cot\phi+(\sigma'_1-\sigma'_3)/2}$$

$$or \ \sigma'_1(1-\sin\phi) = 2c \ \cos\phi+\sigma'_3(1+\sin\phi)$$

$$\sigma'_1 = \sigma_3\frac{1+\sin\phi}{1-\sin\phi}+\frac{2c \ \cos\phi}{1-\sin\phi}$$

$$\sigma'_1 = \sigma_3\left(45° +\frac{\phi}{2}\right)+2c \ \tan\left(45° +\frac{\phi}{2}\right)$$

Note that the plane of failure makes an angle of $45° +\phi/2$ with the major principal plane.

If a clay s initially consolidated by an encompassing chamber pressure of $\sigma_c = \sigma'_c$ and allowed to swell under a reduced chamber pressure of $\sigma_3 = \sigma'_3$ the specimen will be overconsolidated. The failure envelope obtained from consolidated drained triaxial tests of these types of specimens has two distinct branches, as shown in Figure. Portion *ab* of the failure envelope has a flatter slope with a cohesion intercept, and the portion *bc* represents a normally consolidated stage following the equation $s = \sigma' \tan\phi_{bc}$.

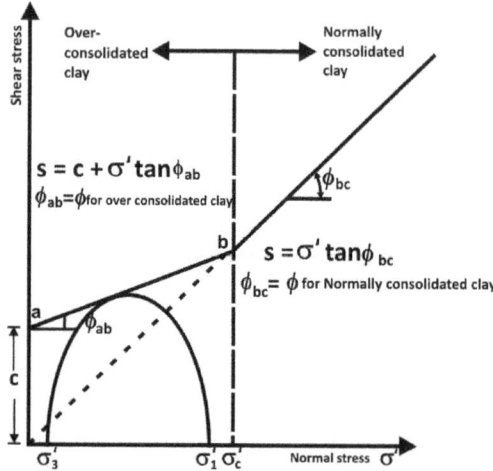

Failure envelope of a clay with perconsoldation pressure $= \sigma'_c$

The shear strength of clays at very large strains is referred to as residual shear strength (i.e., the ultimate shear strength). It has been proved that the residual strength of a given soil is independent of past stress history

$$s_{residual} = \sigma' \tan \phi_{ult}$$

(i.e., the c components is o). For triaxial tests,

$$\phi_{ult} = \sin^{-1}\left(\frac{\sigma'_1 - \sigma'_3}{\sigma'_1 + \sigma'_3}\right)_{residual}$$

Where $\sigma'_1 = \sigma'_3 + \Delta\sigma_{ult}$

The residual friction angle in clays is of importance in subjects such as the long-term stability of slopes.

> ➢ Consolidated undrained test

In the consolidated undrained test, the soil specimen is first consolidated by a chamber confining pressure $\sigma_3$; full drainage from the specimen is allowed. After complete dissipation of excess pore water pressure, $u_c$, generated by the confining pressure, the deviator stress $\Delta\sigma$ is increased to cause failure of the specimen. During this phase of loading, the drainage line from the specimen is closed. Since drainage is not permitted, the pore water pressure (pore water pressure due to deviator stress, $u_d$) in the specimen increases. Simultaneous measurements of $\Delta\sigma$ and $u_d$ are made during the test. Figure shows the nature of the variation of $\Delta\sigma$ and $u_d$ with axial strain; also shown is the nature of the variation of the pore water pressure parameter $A[A = u_d / \Delta\sigma]$ with axial strain. The value of A at failure, $A_f$ is positive for normally consolidated clays and becomes negative for overconsolidated clays. Thus, $A_f$ is dependent on the overconsolidated ratio. The overconsolidation ratio, OCR, for triaxial test conditions may be defined as

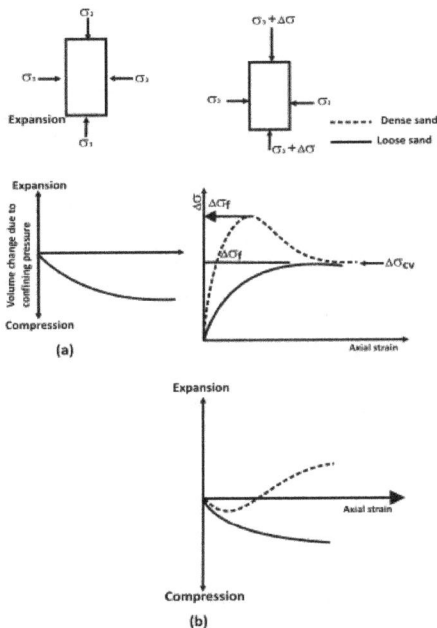

Consolidation undrained triaxial test.
(a) Application of confining pressure (b) application of deviator stress

$$OCR = \frac{\sigma'_c}{\sigma_3}$$

Where $\sigma'_c = \sigma_c$ is the maximum chamber pressure at which the specimen is consolidated and then allowed to rebound under a chamber pressure of $\sigma_3$.

At failure,

$$\text{total major principal stress} = \sigma_1 = \sigma_3 + \Delta\sigma_f$$

$$\text{total minor principal stress} = \sigma_3$$

$$\text{pore water pressure at failure} = u_{d(failure)} = A_f \Delta\sigma_f$$

$$\text{effective major principal stress} = \sigma_1 - A_f \Delta\sigma_f = \sigma_1$$

$$\text{effective minor principal stress} = \sigma_3 - A_f \Delta\sigma_f = \sigma_3$$

consolidated undrained tests on a number of specimens can be conducted to determine the shear strength parameters of a soil, as shown for the case of a normally consolidated clay in Figure. The total-stress Mohr's circles (circles A and B) for two tests are shown by the broken lines. The effective-stress Mohr's circles C and D correspond to the total-stress circles Ai and B, respectively. Since C and D are effective- stress circles at failure, a common tangent drawn to these circles will give the Mohr-Coulomb failure envelope given by the equation $s = \sigma' \tan \phi$. If we draw a common tangent to the total-stress circles, it will be a straight line passing through the origin. This is the total-stress failure envelope, and it may be given by

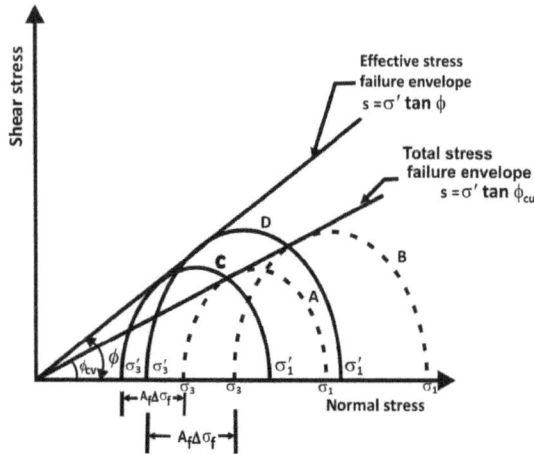

Consolidated undrained test results-normally consolidated clay

$$s = \sigma \tan \phi_{cu}$$

Where $\phi_{cu}$ is the consolidated undrained angle of friction.

The total-stress failure envelope for an over consolidated clay will be of the nature shown in Figure and can be given by the relation

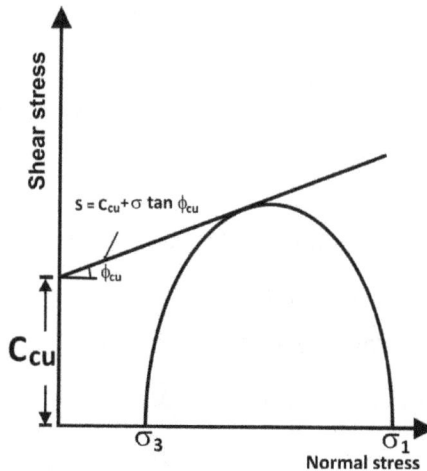

Consolidated undrained test-total stress envelope for overconsolidated clay

$$s = C_{cu} + \sigma \tan \phi_{cu}$$

Where $C_{cu}$ is the intercept of the total-stress failure envelope along the shear stress axis.

The shear strength parameters for overconsolidated clay based on effective stress, i. e., $c$ and $\phi$ can be obtained by plotting the effective-stress Mohr's circle and then drawing a common tangent to $c$ and $\phi$.

As in consolidated drained tests, shear failure in the specimen can be produced by axial compression or extension by changing the loading conditions.

> Unconsolidated undrained test

In unconsolidated undrained triaxial tests, drainage from the specimen is not allowed at any stage. First, the chamber confining pressure $\sigma_3$ is applied, after which the deviator stress $\Delta\sigma$ is increased until failure occurs.

For these tests.

Total major principal stress $= \sigma_3 + \Delta\sigma_f = \sigma_1$

Total minor principal stress $= \sigma_3$

Tests of this type can be performed quickly since drainage is not allowed. For a saturated soil, the deviator stress at failure, $\Delta\sigma_f$ is practically the same irrespective of the confining pressure $\sigma_3$. So, the total-stress failure envelope can be assumed to be a horizontal line, and $\phi = 0$. The undrained shear strength can be expressed as

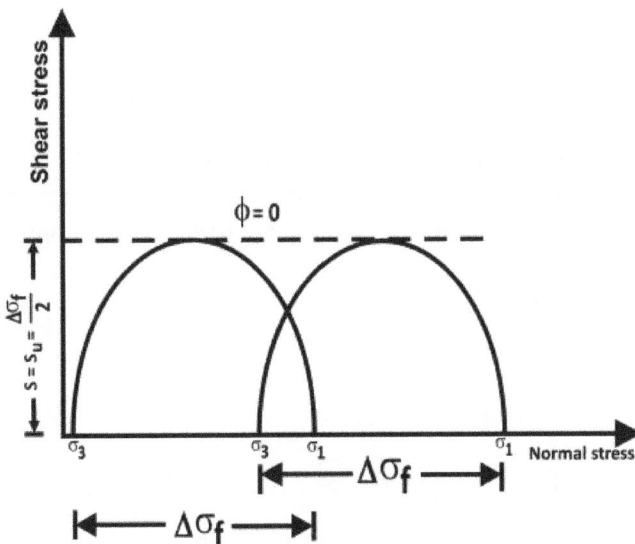

Unconsolidated undrained triaxial test

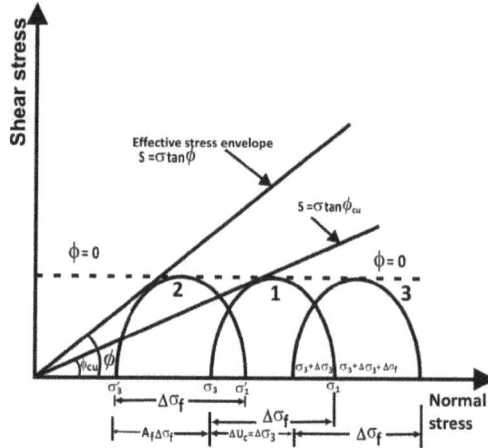

$$s = S_u = \frac{\Delta\sigma_f}{2}$$

This generally referred to as the shear strength based on $\phi = 0$ concept.

The fact that the strength of saturated clay sin unconsolidated undrained loading conditions is the same irrespective of the confining pressure $\sigma_3$ can be explained with the help of Figure. If a saturated clay specimen A is consolidated under a chamber confining pressure of $\sigma_3$ and then sheared to failure under undrained conditions, the Mohr's circle at failure will be represented by circle no 1. The effective-stress Mohr's circle corresponding to circle no 1 is circle no. 2, which touches the effective-stress failure envelope. If a similar soil specimen B, consolidated under a chamber confining pressure of $\sigma_3$ is subjected to an additional confining pressure of $\Delta\sigma_3$ without allowing drainage, the pre water pressure will increase by $\Delta u_c$.

Also $\Delta u_c = B\Delta\sigma_3$ and, for saturated soils, $= 1, so, \Delta u_c = \Delta\sigma_3$.

Since the effective confining pressure of specimen B is the same as specimen A, it will fall with the same deviator stress, $\Delta\sigma_f$. The total-stress Mohr's circle for this specimen (i.e., B) at failure can be given by circle no. 3. So, at failure, for specimen B.

Total minor principal stress $= \sigma_3 + \Delta\sigma_3$

Total major principal stress $= \sigma_3 + \Delta\sigma_3 + \Delta\sigma_f$

The effective stresses for the specimen are as follows:

Effective major principal stress $= \left(\sigma_3 + \Delta\sigma_3 + \Delta\sigma_f\right) - \left(\Delta u_c + A_f\Delta\sigma_f\right)$

$$= \sigma_3 + \Delta\sigma_f) - A_f\Delta\sigma_f$$

$$= \sigma_1 - A_f\Delta\sigma_f = \sigma'_1$$

Effective minor principal stress $= \left( \sigma_3 + \Delta\sigma_3 \right) - \left( \Delta u_c + A_f \Delta\sigma_f \right)$

$$= \sigma_3 - A_f \Delta\sigma_f = \sigma'_3$$

The above principal stresses are the same as those we had for specimen A. Thus, the effective-stress Mohr's circle at failure for specimen B will be the same as that for specimen A, i.e., circle no 1.

The value of $\Delta\sigma_3$ could be of any magnitude in specimen B; in all cases, $\Delta\sigma_f$ would be the same

Example 1: Consolidated drained triaxial tests on two specimens of a soil gave the following results:

| Test no. | Confining pressure $\sigma_3, kN / m^2$ | Deviator stress at failure $\Delta\sigma_f, kN / m^2$ |
|---|---|---|
| 1 | 70 | 440.4 |
| 2 | 92 | 474.7 |

Determine the values of c and $\phi$ for the soil.

Solution: From equation, $\sigma_1 = \sigma_3 \tan^2 \left( 45° + \phi / 2 \right) + 2c \tan \left( 45° + \phi / 2 \right)$. For test 1, $\sigma_3 = 70 kN / m^2$; $\sigma_1 = \sigma_3 + \Delta\sigma_f = 70 + 440.4 = 510.4 kN / m^2$. So,

$$510.4 = 70 \tan^2 \left( 45° + \frac{\phi}{2} \right) + 2c \tan \left( 45° + \frac{\phi}{2} \right) \qquad (a)$$

Similarly, for test 2, $\sigma_3 = 92 kN / m^2$; $\sigma_1 = 92 + 474.7 = 566.7 kN / m^2$. Thus,

$$566.7 = 92 \tan^2 \left( 45° + \frac{\phi}{2} \right) + 2c \tan \left( 45° + \frac{\phi}{2} \right) \qquad (b)$$

Subtracting equation (a) from (b),

$$56.3 = 22 \tan^2 \left( 45° + \frac{\phi}{2} \right)$$

$$\phi = 2 [\tan^{-1} \left( \frac{56.3}{22} \right)^{1/2} - 45°] = 26°$$

Substituting $\phi = 26°$ in equation (a)

$$c = \frac{510.4 - 70\tan^2\left(45^\circ + 26/2\right)}{2\tan\left(45^\circ + 26/2\right)} = \frac{510.4 - 70(2.56)}{2(1.6)} = 103.5 kN/m^2$$

Example 2: A normally consolidated clay specimen was subjected to a consolidated undrained test. At failure, $\sigma_3 = 100 kN/m^2, \sigma_1 = 204 kN/m^2,$ and $u_d = 50 kN/m^2$. Determine $\phi_{cu}$ and $\phi$.

Solution: Referring to Figure,

$$\sin\phi_{cu} = \frac{\overline{ab}}{\overline{Oa}} = \frac{(\sigma_1 - \sigma_3)/2}{(\sigma_1 + \sigma_3)/2} = \frac{\sigma_1 - \sigma_3}{\sigma_1 + \sigma_3} = \frac{204 - 100}{204 + 100} = \frac{96}{304}$$

Hence

$$\phi_{cu} = 18.9^\circ$$

Again,

$$\sin\phi = \frac{\overline{cd}}{\overline{Oc}} = \frac{\sigma'_1 - \sigma'_3}{\sigma'_1 - \sigma'_3}$$

$$\sigma'_3 = 100 - 50 = 50 kN/m^2$$

$$\sigma'_1 = 204 - 50 = 154 kN/m^2$$

So,

$$\sin\phi = \frac{154 - 50}{154 + 50} = \frac{104}{204}$$

Hence

$$\phi = 30.7^\circ$$

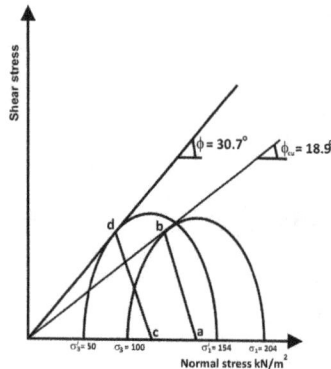

Example 3: For a saturated clay soil, the following are the results of some consolidated drained triaxial tests at failure:

| Test no. | $p' = \dfrac{\sigma'_1 + \sigma'_3}{2}$ $lb/in^2$ | $q' = \dfrac{\sigma'_1 - \sigma'_3}{2}$ $lb/in^2$ |
|---|---|---|
| 1 | 60 | 25.6 |
| 2 | 90 | 36.5 |
| 3 | 110 | 44.0 |
| 4 | 180 | 68.0 |

Draw a $p'$ vs. $q'$ diagram, and from that determine $c$ and $\phi$ for the soil.

Solution: The diagram of $q'$ vs. $p'$ is shown in Figure; this is a straight line, and the equation of it may be written in the form

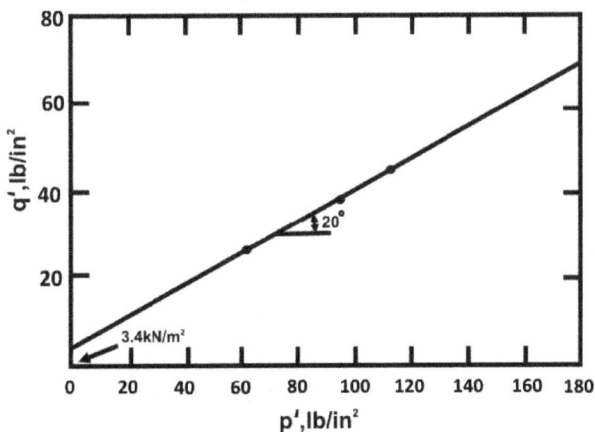

$$q' = m + p' \tan \qquad\qquad (c)$$

Now equation can be written in the form

$$\frac{\sigma'_1 - \sigma'_3}{2} = c \cos \phi + \frac{\sigma'_1 + \sigma'_3}{2} \sin \phi \qquad\qquad (d)$$

Comparing equations (c) and (d) we find $m = c \cos \phi$ or $c = m / \cos \phi$ and $\tan \alpha = \sin \phi$. From Figure, $m = 3.4 lb/in^2$ and $\alpha = 20°$. So,

$$\phi = \sin^{-1}(\tan 20°) = 21.34°$$

$$And\ c = \frac{m}{\cos \alpha} = \frac{3.4}{\cos 21.34°} = 3.65\,lb/in^2$$

# Mohr–Coulomb Failure Criterion

The Mohr–Coulomb failure criterion represents the linear envelope that is obtained from a plot of the shear strength of a material versus the applied normal stress. This relation is expressed as

$$\tau = \sigma \tan(\phi) + c$$

where $\tau$ is the shear strength, $\sigma$ is the normal stress, $c$ is the intercept of the failure envelope with the $\tau$ axis, and $\phi$ is the slope of the failure envelope. The quantity $c$ is often called the cohesion and the angle $\phi$ is called the angle of internal friction. Compression is assumed to be positive in the following discussion. If compression is assumed to be negative then $\sigma$ should be replaced with $-\sigma$.

If $\phi = 0$, the Mohr–Coulomb criterion reduces to the Tresca criterion. On the other hand, if $\phi = 90°$ the Mohr–Coulomb model is equivalent to the Rankine model. Higher values of $\phi$ are not allowed.

From Mohr's circle we have

$$\sigma = \sigma_m - \tau_m \sin\phi \; ; \; \tau = \tau_m \cos\phi$$

where

$$\tau_m = \frac{\sigma_1 - \sigma_3}{2} \; ; \; \sigma_m = \frac{\sigma_1 + \sigma_3}{2}$$

and $\sigma_1$ is the maximum principal stress and $\sigma_3$ is the minimum principal stress.

Therefore, the Mohr–Coulomb criterion may also be expressed as

$$\tau_m = \sigma_m \sin\phi + c \cos\phi \; .$$

This form of the Mohr–Coulomb criterion is applicable to failure on a plane that is parallel to the $\sigma_2$ direction.

## Mohr–Coulomb Failure Criterion in Three Dimensions

The Mohr–Coulomb criterion in three dimensions is often expressed as

$$\begin{cases} \pm\dfrac{\sigma_1 - \sigma_2}{2} = \left[\dfrac{\sigma_1 + \sigma_2}{2}\right] sin(\phi) + c\,cos(\phi) \\[2em] \pm\dfrac{\sigma_2 - \sigma_3}{2} = \left[\dfrac{\sigma_2 + \sigma_3}{2}\right] sin(\phi) + c\,cos(\phi) \end{cases}$$

$$\left| \pm \frac{\sigma_3 - \sigma_1}{2} \right| = \left[ \frac{\sigma_3 + \sigma_1}{2} \right] sin(\phi) + c\, cos(\phi).$$

The Mohr–Coulomb failure surface is a cone with a hexagonal cross section in deviatoric stress space.

The expressions for $\tau$ and $\sigma$ can be generalized to three dimensions by developing expressions for the normal stress and the resolved shear stress on a plane of arbitrary orientation with respect to the coordinate axes (basis vectors). If the unit normal to the plane of interest is

$$\mathbf{n} = n_1\, \mathbf{e}_1 + n_2\, \mathbf{e}_2 + n_3\, \mathbf{e}_3$$

where $\mathbf{e}_i$, $i = 1,2,3$ are three orthonormal unit basis vectors, and if the principal stresses $\sigma_1, \sigma_2, \sigma_3$ are aligned with the basis vectors $\mathbf{e}_1, \mathbf{e}_2, \mathbf{e}_3$, then the expressions for $\sigma, \tau$ are

$$\sigma = n_1^2 \sigma_1 + n_2^2 \sigma_2 + n_3^2 \sigma_3$$
$$\tau = \sqrt{(n_1\sigma_1)^2 + (n_2\sigma_2)^2 + (n_3\sigma_3)^2 - \sigma^2}$$
$$= \sqrt{n_1^2 n_2^2 (\sigma_1 - \sigma_2)^2 + n_2^2 n_3^2 (\sigma_2 - \sigma_3)^2 + n_3^2 n_1^2 (\sigma_3 - \sigma_1)^2}.$$

The Mohr–Coulomb failure criterion can then be evaluated using the usual expression

$$\tau = \sigma\, tan(\phi) + c$$

for the six planes of maximum shear stress.

## Derivation of Normal and Shear Stress on a Plane

Let the unit normal to the plane of interest be

$$\mathbf{n} = n_1\, \mathbf{e}_1 + n_2\, \mathbf{e}_2 + n_3\, \mathbf{e}_3$$

where $\mathbf{e}_i$, $i = 1,2,3$ are three orthonormal unit basis vectors. Then the traction vector on the plane is given by

$$\mathbf{t} = n_i\, \sigma_{ij}\, \mathbf{e}_j \quad \text{(repeated indices indicate summation)}$$

The magnitude of the traction vector is given by

$$|\mathbf{t}| = \sqrt{(n_j\, \sigma_{1j})^2 + (n_k\, \sigma_{2k})^2 + (n_l\, \sigma_{3l})^2} \quad \text{(repeated indices indicate summation)}$$

Then the magnitude of the stress normal to the plane is given by

$$\sigma = \mathbf{t} \cdot \mathbf{n} = n_i \, \sigma_{ij} \, n_j \quad \text{(repeated indices indicate summation)}$$

The magnitude of the resolved shear stress on the plane is given by

$$\tau = \sqrt{|\mathbf{t}|^2 - \sigma^2}$$

In terms of components, we have

$$\sigma = n_1^2 \sigma_{11} + n_2^2 \sigma_{22} + n_3^2 \sigma_{33} + 2(n_1 n_2 \sigma_{12} + n_2 n_3 \sigma_{23} + n_3 n_1 \sigma_{31})$$

$$\tau = \sqrt{(n_1 \sigma_{11} + n_2 \sigma_{12} + n_3 \sigma_{31})^2 + (n_1 \sigma_{12} + n_2 \sigma_{22} + n_3 \sigma_{23})^2 + (n_1 \sigma_{31} + n_2 \sigma_{23} + n_3 \sigma_{33})^2 - \sigma^2}$$

If the principal stresses $\sigma_1, \sigma_2, \sigma_3$ are aligned with the basis vectors $\mathbf{e}_1, \mathbf{e}_2, \mathbf{e}_3$, then the expressions for $\sigma, \tau$ are

$$\sigma = n_1^2 \sigma_1 + n_2^2 \sigma_2 + n_3^2 \sigma_3$$

$$\tau = \sqrt{(n_1 \sigma_1)^2 + (n_2 \sigma_2)^2 + (n_3 \sigma_3)^2 - \sigma^2}$$

$$= \sqrt{n_1^2 n_2^2 (\sigma_1 - \sigma_2)^2 + n_2^2 n_3^2 (\sigma_2 - \sigma_3)^2 + n_3^2 n_1^2 (\sigma_3 - \sigma_1)^2}$$

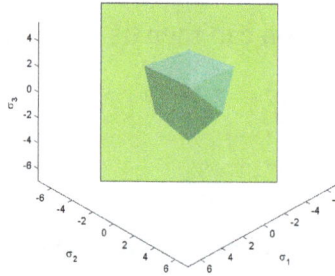

Mohr–Coulomb yield surface in the $\pi$-plane for $c = 2, \phi = 20^\circ$

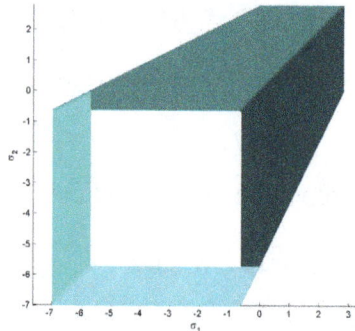

Trace of the Mohr–Coulomb yield
surface in the $\sigma_1 - \sigma_2$-plane for $c = 2, \phi = 20^\circ$

In 1910, Mohr presented a theory for rupture in materials. The failure along a plane in a material occurs by a critical combination of normal and shear stresses, and not by normal or shear stress alone. The functional relation between normal and shear stress on the failure plane can be given by

$$s = f(\sigma)$$

Where s is the shear stress at failure and $\sigma$ is the normal stress on the failure plane. The failure envelope defined by equation is shown in Figure.

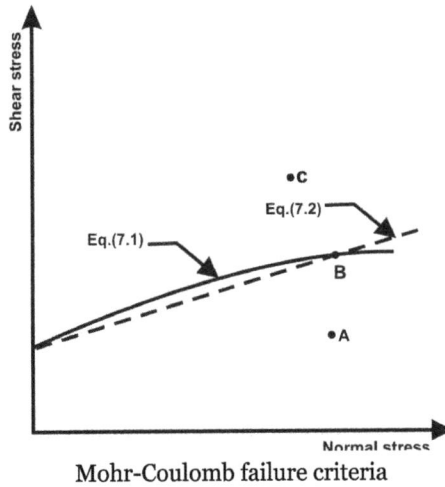

Mohr-Coulomb failure criteria

In 1776, Coulomb defined the function $f(\sigma)$ as

$$s = c + \sigma \tan \phi$$

Where c is cohesion and $\phi$ is the angle of friction of the soil.

Equation is generally referred to as the Mohr-Coulomb failure criteria. If the normal and shear stresses on a plane in a soil mass are such that they plot as point A, shear failure will not occur along that plane.

Shear failure along a plane will occur if the stresses plot as point B, which falls on the failure envelope. A state of stress plotting as point C cannot exits, since this falls above the failure envelope; shear failure would have occurred before this condition was reached.

In saturated soils, the stress carried by the soil solids is the effective stress and so equation must be modified:

$$s = c + (\sigma - u) \tan \phi = c + \sigma' \tan \phi$$

Where u is the pore water pressure and $\sigma'$ is the effective stress on the plane.

The term $\phi$ is also referred to as the drained friction angle. For sand, inorganic silts, and normally consolidated clays, $c \approx 0$. The value of c is greater than zero for over consolidated clays.

# Critical State Soil Mechanics

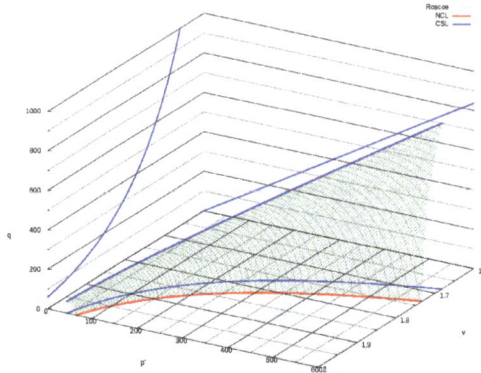

Normally consolidated soil goes to critical state along the stress path on Roscoe surface

Critical State Soil Mechanics is the area of soil mechanics that encompasses the conceptual models that represent the mechanical behavior of saturated remolded soils based on the *Critical State* concept.

## Formulation

The Critical State concept is an idealization of the observed behavior of saturated remoulded clays in triaxial compression tests, and it is assumed to apply to undisturbed soils. It states that soils and other granular materials, if continuously distorted (sheared) until they flow as a frictional fluid, will come into a well-defined critical state. At the onset of the critical state, shear distortions $\varepsilon_s$ occur without any further changes in mean effective stress $p'$, deviatoric stress $q$ (or yield stress, $\sigma_y$, in uniaxial tension according to the von Mises yielding criterion), or specific volume $v$:

$$\frac{\partial p'}{\partial \varepsilon_s} = \frac{\partial q}{\partial \varepsilon_s} = \frac{\partial v}{\partial \varepsilon_s} = 0$$

where,

$$v = 1 + e$$

$$p' = \frac{1}{3}(\sigma'_1 + \sigma'_2 + \sigma'_3)$$

$$q = \sqrt{\frac{(\sigma'_1 - \sigma'_2)^2 + (\sigma'_2 - \sigma'_3)^2 + (\sigma'_1 - \sigma'_3)^2}{2}}$$

However, for triaxial conditions $\sigma'_2 = \sigma'_3$. Thus,

$$p' = \frac{1}{3}(\sigma'_1 + 2\sigma'_3)$$

$$q = (\sigma'_1 - \sigma'_3)$$

All critical states, for a given soil, form a unique line called the *Critical State Line* (*CSL*) defined by the following equations in the space $(p', q, v)$:

$$q = Mp'$$

$$v = \Gamma - \lambda \ln(p')$$

where $M$, $\Gamma$, and $\lambda$ are soil constants. The first equation determines the magnitude of the deviatoric stress $q$ needed to keep the soil flowing continuously as the product of a frictional constant $M$ (capital $\mu$) and the mean effective stress $p'$. The second equation states that the specific volume $v$ occupied by unit volume of flowing particles will decrease as the logarithm of the mean effective stress increases.

## History

In an attempt to advance soil testing techniques, Kenneth Harry Roscoe of Cambridge University, in the late forties and early fifties, developed a simple shear apparatus in which his successive students attempted to study the changes in conditions in the shear zone both in sand and in clay soils. In 1958 a study of the yielding of soil based on some Cambridge data of the simple shear apparatus tests, and on much more extensive data of triaxial tests at Imperial College London from research led by Professor Sir Alec Skempton at the Imperial Geotechnical Laboratories, led to the publication of the critical state concept (Roscoe, Schofield & Wroth 1958).

Roscoe obtained his undergraduate degree in mechanical engineering and his experiences trying to create tunnels to escape when held as a prisoner of war by the Nazis during WWII introduced him to soil mechanics. Subsequent to this 1958 paper, concepts of plasticity were introduced by Schofield and publa classic text book (Schofield & Wroth 1968). Schofield was taught at Cambridge by Prof. John Baker, a structural engineer who was a strong believer in designing structures that would fail "plastically". Prof. Baker's theories strongly influenced Schofield's thinking on soil shear. Prof. Baker's views were developed from his pre-war work on steel structures and further informed by his wartime experiences assessing blast-damaged structures and with the design of the "Morrison Shelter", an air-raid shelter which could be located indoors (Schofield 2006).

## Original Cam-Clay Model

The Original Cam-Clay model is based on the assumption that the soil is isotropic, elasto-plastic, deforms as a continuum, and it is not affected by creep. The yield surface of the Cam clay model is described by the equation

$$f(p, q, p_c) = q + M \, p \ln\left[\frac{p}{p_c}\right] \le 0$$

where $q$ is the equivalent stress, $p$ is the pressure, $p_c$ is the pre-consolidation pressure, and $M$ is the slope of the critical state line in $p-q$ space.

The pre-consolidation pressure evolves as the void ratio ($e$) (and therefore the specific volume $v$) of the soil changes. A commonly used relation is

$$e = e_0 - \lambda \ln\left[\frac{p_c}{p_{c0}}\right]$$

where $\lambda$ is the virgin compression index of the soil. A limitation of this model is the possibility of negative specific volumes at realistic values of stress.

An improvement to the above model for $p_c$ is the bilogarithmic form

$$\ln\left[\frac{1+e}{1+e_0}\right] = \ln\left[\frac{v}{v_0}\right] = -\tilde{\lambda} \ln\left[\frac{p_c}{p_{c0}}\right]$$

where $\tilde{\lambda}$ is the appropriate compressibility index of the soil.

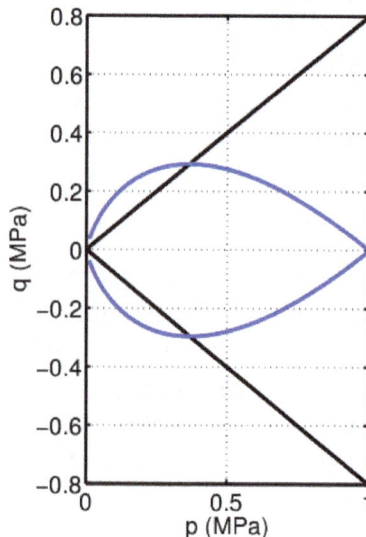

Cam-clay yield surface in p-q space.

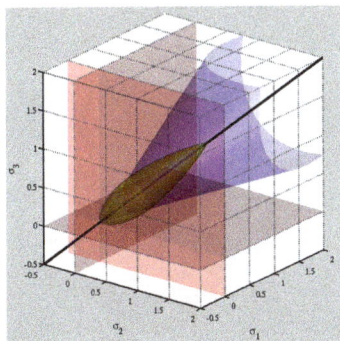

Cam-clay yield surface in principal stress space.

## Modified Cam-Clay Model

Professor John Burland of Imperial College who worked with Professor Roscoe is credited with the development of the modified version of the original model. The difference between the Cam Clay and the Modified Cam Clay (MCC) is that the yield surface of the MCC is described by an ellipse and therefore the plastic strain increment vector (which is perpendicular to the yield surface) for the largest value of the mean effective stress is horizontal, and hence no incremental deviatoric plastic strain takes place for a change in mean effective stress (for purely hydrostatic states of stress). This is very convenient for constitutive modelling in numerical analysis, especially finite element analysis, where numerical stability issues are important (as a curve needs to be continuous in order to be differentiable).

The yield surface of the modified Cam-clay model has the form

$$f(p,q,p_c) = \left[\frac{q}{M}\right]^2 + p(p - p_c) \leq 0$$

where $p$ is the pressure, $q$ is the equivalent stress, $p_c$ is the pre-consolidation pressure, and $M$ is the slope of the critical state line.

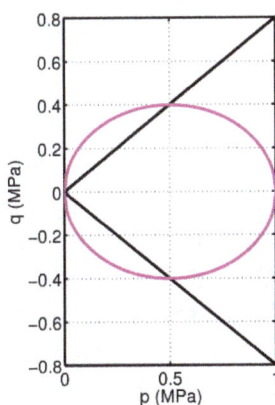

Modified Cam-clay yield surface in p-q space.

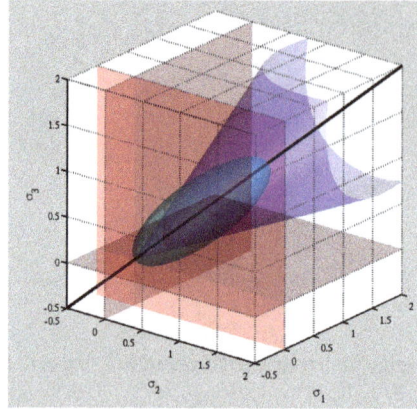

Modified Cam-clay yield surface in principal stress space.

## Critique

The basic concepts of the elasto-plastic approach were first proposed by two mathematicians Daniel C. Drucker and William Prager (Drucker and Prager, 1952) in a short eight page note. In their note, Drucker and Prager also demonstrated how to use their approach to calculate the critical height of a vertical bank using either a plane or a log spiral failure surface. Their yield criterion is today called the Drucker-Prager yield criterion. Their approach was subsequently extended by Kenneth H. Roscoe and others in the soil mechanics department of Cambridge University.

Critical state and elasto-plastic soil mechanics have been the subject of criticism ever since they were first introduced. The key factor driving the criticism is primarily the implicit assumption that soils are made of isotropic point particles. Real soils are composed of finite size particles with anisotropic properties that strongly determine observed behavior. Consequently, models based on a metals based theory of plasticity are not able to model behavior of soils that is a result of anisotropic particle properties, one example of which is the drop in shear strengths post peak strength, i.e., strain-softening behavior. Because of this elasto-plastic soil models are only able to model "simple stress-strain curves" such as that from isotropic normally or lightly over consolidated "fat" clays, i.e., CL-ML type soils constituted of very fine grained particles.

Also, in general, volume change is governed by considerations from elasticity and, this assumption being largely untrue for real soils, results in very poor matches of these models to volume changes or pore pressure changes. Further, elasto-plastic models describe the entire element as a whole and not specifically conditions directly on the failure plane, as a consequence of which, they do not model the stress-strain curve post failure, particularly for soils that exhibit strain-softening post peak. Finally, most models separate out the effects of hydrostatic stress and shear stress, with each assumed to cause only volume change and shear change respectively. In reality, soil structure, being analogous to a "house of cards," shows both shear deformations on the application of pure compression, and volume changes on the application of pure shear.

Additional criticisms are that the theory is "only descriptive," i.e., only describes known behavior and lacking the ability to either explain or predict standard soil behaviors such as, why the void ratio in a one dimensional compression test varies linearly with the logarithm of the vertical effective stress. This behavior, critical state soil mechanics simply assumes as given.

For these reasons, critical-state and elasto-plastic soil mechanics have been subject to charges of scholasticism; the tests to demonstrated its validity are usually "conformation tests" where only simple stress-strain curves are demonstrated to be modeled satisfactorily. In the 1960's and 1970's, Prof. Alan Bishop at Imperial College used to routinely demonstrate the inability of these theories to match the stress-strain curves of real soils. Joseph (2013) has suggested that critical-state and elasto-plastic soil mechanics meet the criterion of a "degenerate research program" a concept proposed by the philosopher of science Imre Lakatos, for theories where excuses are used to justify an inability of theory to match empirical data.

# Shearing Strength of Granular Soils

For granular soils with $c = 0$,

$$s = \sigma' \tan \phi$$

The determination of friction angle $\phi$ is commonly accomplished by one of two methods; the direct shear test or the triaxial test. The test procedures are given below.

## Direct Shear Test

A direct shear test is a laboratory or field test used by geotechnical engineers to measure the shear strength properties of soil or rock material, or of discontinuities in soil or rock masses.

The U.S. and U.K. standards defining how the test should be performed are ASTM D 3080, AASHTO T236 and BS 1377-7:1990, respectively. For rock the test is generally restricted to rock with (very) low shear strength. The test is, however, standard practice to establish the shear strength properties of discontinuities in rock.

The test is performed on three or four specimens from a relatively undisturbed soil sample. A specimen is placed in a *shear box* which has two stacked rings to hold the sample; the contact between the two rings is at approximately the mid-height of the sample. A *confining stress* is applied vertically to the specimen, and the upper ring is pulled laterally until the sample fails, or through a specified strain. The load applied and the strain induced is recorded at frequent intervals to determine a stress–strain curve for each confining stress. Several specimens are tested at varying confining stresses to determine the shear strength parameters, the soil cohesion (c) and the angle of internal friction, commonly known as *friction angle* ($\phi$). The results of the tests on each specimen are plotted on a graph with the peak (or residual) stress on the y-axis and the confining stress on the x-axis. The y-intercept of the curve which fits the test results is the cohesion, and the slope of the line or curve is the friction angle.

Direct shear tests can be performed under several conditions. The sample is normally saturated before the test is run, but can be run at the in-situ moisture content. The rate of strain can be varied to create a test of *undrained* or *drained* conditions, depending whether the strain is applied slowly enough for water in the sample to prevent pore-water pressure buildup. Direct shear test machine is required to perform the test. The test using the direct shear machine determinates the consolidated drained shear strength of a soil material in direct shear.

The advantages of the direct shear test over other shear tests are the simplicity of setup and equipment used, and the ability to test under differing saturation, drainage, and consolidation conditions. These advantages have to be weighed against the difficulty of measuring pore-water pressure when testing in undrained conditions, and possible spuriously high results from forcing the failure plane to occur in a specific location.

The test equipment and procedures are slightly different for test on discontinuities.

A schematic diagram of the direct shear test equipment is shown in Figure. Basically, the test equipment consists of a metal shear box into which the soil specimen is placed. The specimen can be square or circular in plan, about 3 to $4\,in^2$ (19.35 to 25.80 $cm^2$) in area, and about 1 in (25.4 mm) in height. The box is split horizontally into two halves. Normal force on the specimen is applied from the top of the shear box by dead weights. The normal stress on the specimens obtained by the application of dead weights can be as high as $1035\,kN/m^2$. Shear force is applied to the side of the top half of the box to cause failure in the soil specimen. (The two porous stones shown in Figure are not required for tests on dry soil). During the test, the shear displacement of the top half of the box and the change in specimen thickness are recorded by the use of horizontal and vertical dial gauges.

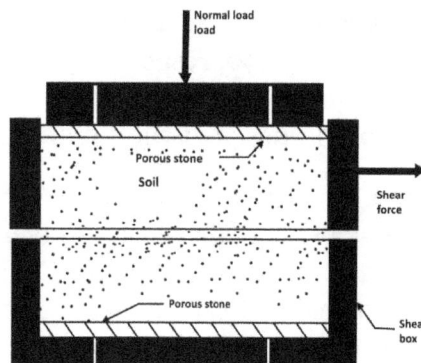

Direct shear test arrangement

Figure Shows the nature of the results of typical direct shear tests in loose, medium, and dense sands. The following observations can be drawn

1. In dense and medium sands, shear stress increases with shear displacement to a maximum or peak value $\tau_m$ and then decreases to an approximately constant value $\tau_{cv}$ at large shear displacement. This constant stress $\tau_{cv}$ is the ultimate shear stress.

2. For loose sands, the shear stress increases with shear displacement to a maximum value and then remains constant.

3. For dense and medium sands, the volume of the specimen initially decreases and then increases with shear displacement. At large values of shear displacement, the volume of the specimen remains approximately constant.

4. For loose sands, the volume of the specimen gradually decreases to a certain value and remains approximately constant thereafter.

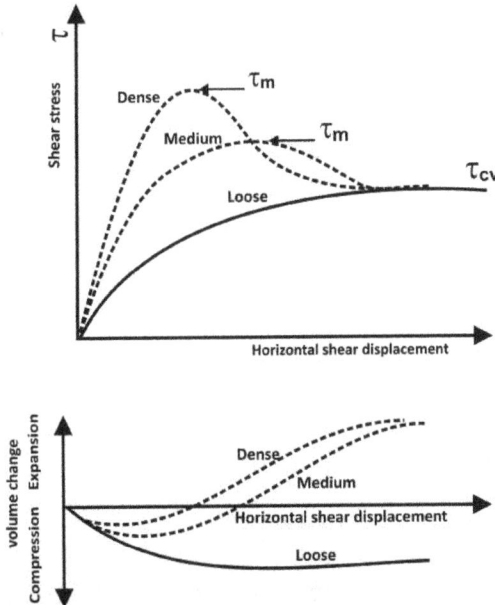

Direct shear test results in loose, medium and dense sands

If dry sand is used for the test, the pore water pressure u is equal to zero, and so the total normal stress $\sigma$ is equal to the effective stress $\sigma'$. The test may be repeated for several normal stresses. The angle of friction $\phi$ for the sand can be determined by plotting a graph of the maximum or peak shear stresses vs. the corresponding normal stresses, as shown in Figure. The Mohr-Coulomb failure envelope can be determined by drawing a straight line through the origin and the points representing the experimental results. The slope of this line will give the peak friction angle $\phi$ of the soil. Similarly, the ultimate friction angle $\phi_{cv}$ can be determined by plotting the ultimate shear stresses $\tau_{cv}$ vs. The corresponding normal stresses, as shown in Figure. The ultimate friction angle $\phi_{cv}$ represents a condition of shearing at constant volume of the specimen.

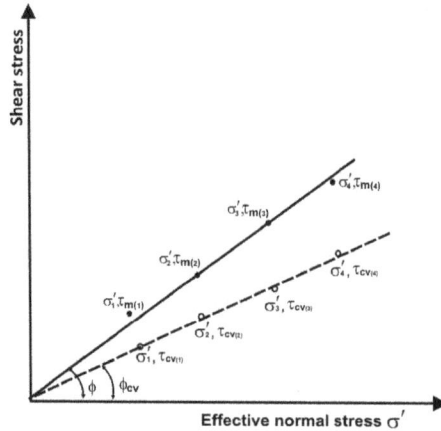

Determination of peak and ultimate friction angle from direct shear test

Some typical values of $\phi \, and \, \phi_{cv}$ for granular soils are given table.

Table: Typical values of $\phi \, and \, \phi_{cv}$ for granular soils

| Type of soil | $\phi$, deg | $\phi_{cv}$ deg |
|---|---|---|
| Sand: round grains | | |
| Loose | 28 to30 | |
| Medium | 30 to 35 | 26 to 30 |
| Dense | 35 to 38 | |
| Sand: angular grains | | |
| Loose | 30 to 35 | |
| Medium | 35 to 40 | 30 to 35 |
| Dense | 40 to 45 | |
| Sandy gravel | 34 to 48 | 33 to 36 |

## Triaxial Shear Test

A triaxial shear test is a common method to measure the mechanical properties of many deformable solids, especially soil (e.g., sand, clay) and rock, and other granular materials or powders. There are several variations on the test.

In a triaxial shear test, stress is applied to a sample of the material being tested in a way which results in stresses along one axis being different from the stresses in perpendicular directions. This is typically achieved by placing the sample between two parallel platens which apply stress in one (usually vertical) direction, and applying fluid pressure to the specimen to apply stress in the perpendicular directions. (Testing apparatus which allows application of different levels of stress in each of three orthogonal directions are discussed below, under "True Triaxial test".)

The application of different compressive stresses in the test apparatus causes shear stress to develop in the sample; the loads can be increased and deflections monitored until failure of the sample. During the test, the surrounding fluid is pressurized, and the stress on the platens is increased until the material in the cylinder fails and forms sliding regions within itself, known as shear bands. The geometry of the shearing in a triaxial test typically causes the sample to become shorter while bulging out along the sides. The stress on the platen is then reduced and the water pressure pushes the sides back in, causing the sample to grow taller again. This cycle is usually repeated several times while collecting stress and strain data about the sample. During the test the pore pressures of fluids (e.g., water, oil) or gasses in the sample may be measured using Bishop's pore pressure apparatus.

From the triaxial test data, it is possible to extract fundamental material parameters about the sample, including its angle of shearing resistance, apparent cohesion, and dilatancy angle. These parameters are then used in computer models to predict how the material will behave in a larger-scale engineering application. An example would be to predict the stability of the soil on a slope, whether the slope will collapse or whether the soil will support the shear stresses of the slope and remain in place. Triaxial tests are used along with other tests to make such engineering predictions.

During the shearing, a granular material will typically have a net gain or loss of volume. If it had originally been in a dense state, then it typically gains volume, a characteristic known as Reynolds' dilatancy. If it had originally been in a very loose state, then contraction may occur before the shearing begins or in conjunction with the shearing.

Sometimes, testing of cohesive samples is done with no confining pressure, in an unconfined compression test. This requires much simpler and less expensive apparatus and sample preparation, though the applicability is limited to samples that the sides won't crumble when exposed, and the confining stress being lower than the in-situ stress gives results which may be overly conservative. The compression test performed for concrete strength testing is essentially the same test, on apparatus designed for the larger samples and higher loads typical of concrete testing.

## Test Execution

For soil samples, the specimen is contained in a cylindrical latex sleeve with a flat, circular metal plate or platen closing off the top and bottom ends. This cylinder is placed into a bath of a hydraulic fluid to provide pressure along the sides of the cylinder. The top platen can then be mechanically driven up or down along the axis of the cylinder to squeeze the material. The distance that the upper platen travels is measured as a function of the force required to move it, as the pressure of the surrounding water is carefully controlled. The net change in volume of the material can also be measured by how much water moves in or out of the surrounding bath, but is typically measured - when the sample is saturated with water - by measuring the amount of water that flows into or out of the sample's pores.

## Rock

For testing of high-strength rock, the sleeve may be a thin metal sheeting rather than latex. Triaxial testing on strong rock is fairly seldom done because the high forces and pressures required to break a rock sample require costly and cumbersome testing equipment.

## Effective Stress

The effective stress on the sample can be measured by using a porous surface on one platen, and measuring the pressure of the fluid (usually water) during the test, then calculating the effective stress from the total stress and pore pressure.

## Triaxial Test to Determine the Shear Strength of a Discontinuity

The *triaxial test* can be used to determine the shear strength of a discontinuity. A homogeneous and isotropic sample fails due to shear stresses in the sample. If a sample with a discontinuity is orientated such that the discontinuity is about parallel to the plane in which maximum shear stress will be developed during the test, the sample will fail due to shear displacement along the discontinuity, and hence, the shear strength of a discontinuity can be calculated.

## Types of Triaxial Tests

There are several variations of the triaxial test:

## Consolidated Drained (CD)

In a *consolidated drained* test the sample is consolidated and sheared in compression slowly to allow pore pressures built up by the shearing to dissipate. The rate of axial deformation is kept constant, i.e., is strain controlled. The idea is that the test allows the sample and the pore pressures to fully consolidate (i.e., *adjust*) to the surrounding stresses. The test may take a long time to allow the sample to adjust, in particular low permeability samples need a long time to drain and adjust strain to stress levels.

## Consolidated Undrained (CU)

In a *consolidated undrained* test the sample is not allowed to drain. The shear characteristics are measured under undrained conditions and the sample is assumed to be fully saturated. Measuring the pore pressures in the sample (sometimes called CUpp) allows approximating the consolidated-drained strength.

## Unconsolidated Undrained (UU)

In an *unconsolidated undrained* test the loads are applied quickly, and the sample is not allowed to consolidate during the test. The sample is compressed at a constant rate (*strain-controlled*).

# True Triaxial Test

Triaxial testing systems have been developed to allow independent control of the stress in three perpendicular directions. This allows investigation of stress paths not capable of being generated in axisymmetric triaxial test machines, which can be useful in studies of cemented sands and anisotropic soils. The test cell is cubical, and there are six separate plates applying pressure to the specimen, with LVDTs reading movement of each plate. Pressure in the third direction can be applied using hydrostatic pressure in the test chamber, requiring only 4 stress application assemblies. The apparatus is significantly more complex than for axisymmetric triaxial tests, and is therefore less commonly used.

## Free End Condition in Triaxial Testing

The Danish triaxial in action

Triaxial tests of classical construction had been criticized for their nonuniform stress and strain field imposed within the specimen during larger deformation amplitudes. The highly localized discontinuity within a shear zone is caused by combination of rough end plates and specimen height.

To test specimens during larger deformation amplitude, "new" and "improved" version of the triaxial apparatus were made. Both the "new" and the "improved" triaxial follow the same principle - sample height is reduced down to one diameter height and friction with the end plates is canceled.

The classical apparatus uses rough end plates - the whole surface of the piston head is made up of rough, porous filter. In upgraded apparatuses the tough end plates are replaced with smooth, polished glass, with a small filter at the center. This configuration allows a specimen to slide / expand horizontally while sliding along the polished glass. Thus, the contact zone between sample and the end plates does not buildup unnecessary shear friction, and a linear / isotropic stress field within the specimen is sustained.

Due to extremely uniform, near isotropic stress field - isotropic yielding takes place. During isotropic yielding volumetric strain is isotopically distributed within the specimen, this improves measurement of volumetric response during CD tests and pore water pressure during CU loading. Also, isotropic yielding makes the specimen expand radially in uniform manner, as it is compressed axially. The walls of a cylindrical specimen remain straight and vertical even during large strain amplitudes (50% strain amplitude was documented by Vardoulakis (1980), using "improved" triaxial, on non saturated sand). This is in contrast with classical setup, where the specimen forms a bugle in the center, while keeping a constant radius at the contact with the end plates.

Post-liquefaction testing. The fine sand specimen was liquefied during CU cycles and recovered with CD cycles many times. The wrinkles formed due to extreme volume change imposed by iterating between CU liquefaction and draining. In liquefied state sample become soft enough to imprint thin latex. During CD cycles - stiff enough to preserve the imprinted pattern. No bulging or shear rupture is present despite numerous instances of pure plastic yielding.

The "new" apparatus has been upgraded to "the Danish triaxial" by L.B.Ibsen. The Danish triaxial can be used for testing all soil types. It provides improved measurements of volumetric response - as during isotropic yielding, volumetric strain is distributed isotopically within the specimen. Isotropic volume change is especially important for CU testing, as cavitation of pore water sets the limit of undrained sand strength. Measurement precision is improved by taking measurements near the specimen. The load cell is submerged and in direct contact with the upped pressure head of the specimen. Deformation transducers are attached directly to the piston heads as well. Control of the apparatus is highly automated, thus cyclic loading can be applied with great efficiency and precision.

The combination of high automation, improved sample durability and large deformation compatibility expands the scope of triaxial testing. The Danish triaxial can yield CD and CU sand specimens into plasticity without forming a shear rupture or bulging. A sample can be tested for yielding multiple times in a single, continuous loading sequence. Samples can even be liquefied to a large strain amplitude, then crushed to CU failure. CU tests can be allowed to transition into CD state, and cyclic tested in CD mode to observe post liquefaction recovery of stiffness and strength. This allows to control the specimens to a very high degree, and observe sand response patterns which are not accessible using classical triaxial testing methods.

## Test Standards

The list is not complete; only the main standards are included. For a more extensive listing, please refer to the websites of ASTM International (USA), British Standards (UK), International Organization for Standardization (ISO), or local organisations for standards.

- ASTM D7181-11: Standard Test Method for Consolidated Drained Triaxial Compression Test for Soils

- ASTM D4767-11 (2011): Standard Test Method for Consolidated Undrained Triaxial Compression Test for Cohesive Soils

- ASTM D2850-03a (2007): Standard Test Method for Unconsolidated-Undrained Triaxial Compression Test on Cohesive Soils

- BS 1377-9:1990 Part 8: Shear strength tests (effective stress)Triaxial Compression Test

- ISO/TS 17892-8:2004 Geotechnical investigation and testing—Laboratory testing of soil—Part 8: Unconsolidated undrained triaxial test

- ISO/TS 17892-9:2004 Geotechnical investigation and testing—Laboratory testing of soil—Part 9: Consolidated triaxial compression tests on water-saturated soils

A schematic diagram of a triaxial test equipment is shown in Figure. In this type of test, a soil specimen about 1.5 in (38.1 mm) in diameter and 3 in (76.2 mm) in length is generally used. The specimen is enclosed inside a thin rubber membrane and placed inside a cylindrical plastic chamber. For conducting the test, the chamber is usually filled with water or glycerin. The specimen is subjected to a confining pressure $\sigma_3$ by application of pressure to the fluid in the chamber. (Air can sometimes be used as a medium for applying the confining pressure). Connections to measure drainage into or out of the specimen or pressure in the pore water are provided. To cause shear failure in the soil, an axial stress $\Delta\sigma$ is applied through a vertical loading ram. This is also referred to as deviator stress. For determination of $\phi$ dry or fully saturated soil can be used. If saturated soil is used, the drainage connection is kept open during the application of the confining pressure and the deviator stress. Thus, during the test the excess pore water pressure in the specimen is equal to zero. The volume of the water drained from the specimen during the test provides a measure of the volume change of the specimen.

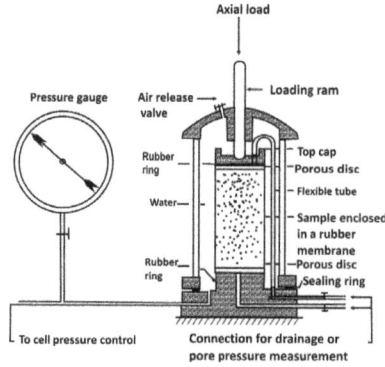

Triaxial test equipment.

For drained test, the total stress is equal to the effective stress Thus, the major effective principal stress is $\sigma'_1 = \sigma_1 = \sigma_3 + \Delta\sigma$; the minor effective principal stress is $\sigma'_3 = \sigma_3$; and the intermediate effective principal stress is $\sigma'_2 = \sigma'_3$.

At failure, the major effective principal stress is equal to $\sigma_3 + \Delta\sigma_f$ where $\Delta\sigma_f$ is the deviator stress at failure, and the minor effective principal stress is $\sigma_3$. Figure shows the nature of the variation of $\Delta\sigma$ with axial strain for loose and dense granular soils. Several tests with similar specimens can be conducted by the using different confining pressure $\sigma_3$. The value of the soil peak friction angle $\phi$ can be determined by plotting effective-stress Mohr's circles for various tests and drawing a common tangent to these Mohr's circles passing through the origin. This is shown in Figure a. The angle that this envelope makes with the normal stress axis is equal to $\phi$. It can be seen from Figure b that

$$\sin\phi = \frac{\overline{ab}}{\overline{Oa}} = \frac{(\sigma'_1 - \sigma'_3)/2}{(\sigma'_1 + \sigma'_3)/2}$$

$$Or\ \phi = \sin^{-1}\frac{(\sigma'_1 - \sigma'_3)}{(\sigma'_1 + \sigma'_3)_{failure}} \qquad\qquad 5$$

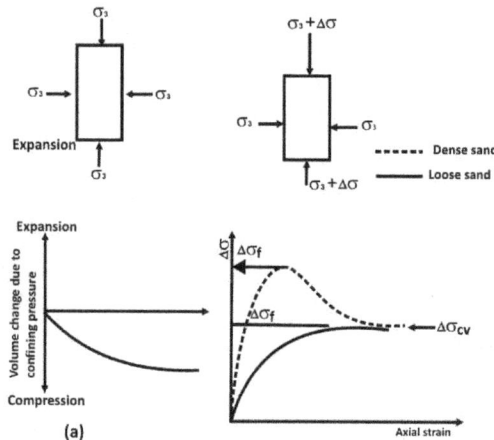

(a)

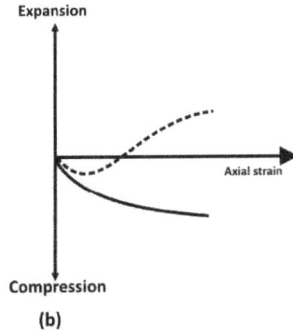

Drained triaxial test in granular soils (a) application of confining pressure
(b) application of deviator stress

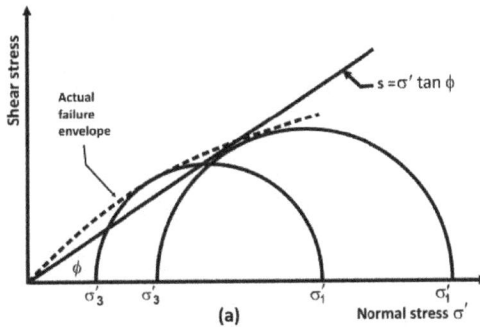

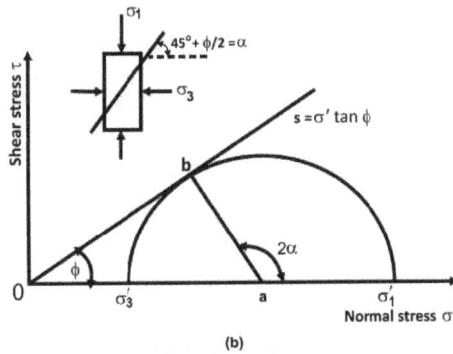

Drained triaxial test results

However, it must be pointed out that in Figure a the failure envelope defined by the equation $s = \sigma' \tan \phi$ is an approximation to the actual curved failure envelope. The ultimate friction angle $\phi_{cv}$ for a given test can be also be determined from the equation

$$\phi_{cv} = \sin^{-1} \left[ \frac{\sigma'_{1(cv)} - \sigma'_3}{\sigma'_{1(cv)} + \sigma'_3} \right]$$

Where $\sigma'_{1(cv)} = \sigma'_3 + \Delta\sigma_{(cv)}$. For similar soils, the friction angle $\phi$ determined by triaxial tests is slightly lower ( o to $3°$ ) than that obtained from direct shear tests.

The axial compression triaxial test described above is the conventional type. However, the loading process on the specimen in a triaxial chamber can be varied in several ways. In general, the tests can be divided into two major groups: axial compression tests and axial extension tests.

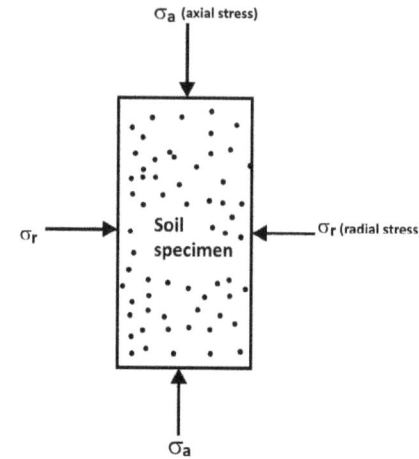

Soil specimen with axial and radial stress

## Shear Strength of Granular Soils under Plane Strain Condition

The results obtained from triaxial tests are widely used for the design of structures. However, under structures such as continuous wall footings the soils are actually subjected to a plane strain type of loading, i.e., the strain in the direction of the intermediate principal stress is equal to zero. Several investigators have attempted to evaluate the effect of plane strain type of loading on the angle of friction of granular soils.

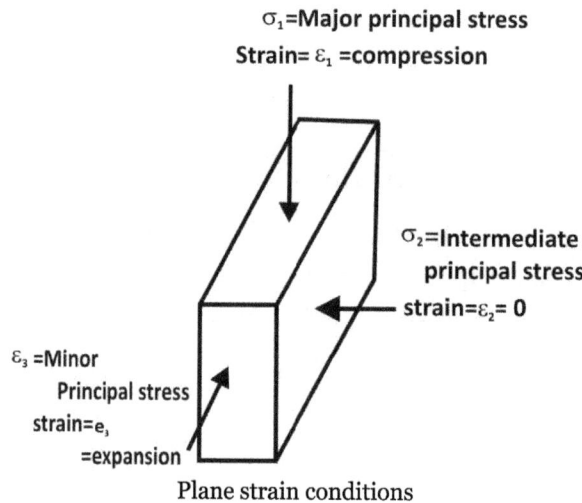

Plane strain conditions

$$\phi_p = drained\ friction\ angle\ obtained\ from\ plane\ strain\ tests$$

$$\phi_t = drained\ friction\ angle\ obtained\ from\ triaxial\ tests$$

Figure shows the results of the initial tangent modulus $\left(E = \Delta\sigma'/\Delta\epsilon_1\right)$ for various confining pressures. For gives values of $\sigma'_3$ the initial tangent modulus for plane strain loading shows a higher value than that for triaxial loading, although in both cases E increases exponentially with the confining pressure.

The variation of Poisson's ratio v with the confining pressure for plane strain and triaxial loading condition is shown in Figure. The values of v were calculated by measuring the change of the volume of specimens and the corresponding axial strains during loading. The derivation of the equation used for finding v can be explained with the aid of Figure. Assuming compressive strain to be positive, for the stresses shown in Figure.

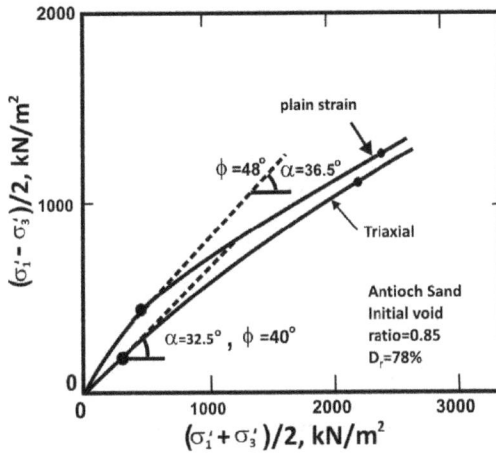

Strength of Antioch sand under drained condition. (Redrawn after K.L. Lee, Comparison of Plane Strain and Triaxial Tests on Sand, J. Soil Mech. Found. Div., ASCE, vol. 96, no. SM3, 1970)

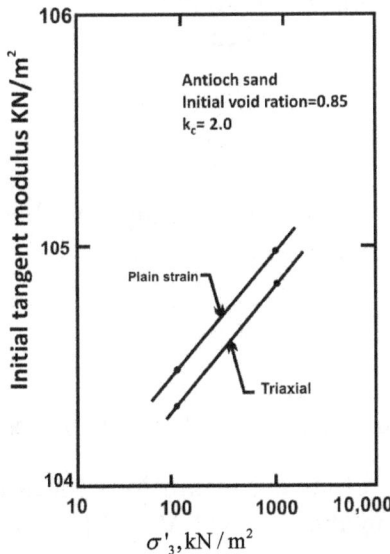

Initial tangent modulus from drained tests on Antioch sand.
(Redrawn after K.L. Lee, Comparison of Plane Strain and Triaxial Tests on Sand,
J. Soil Mech. Found. Div., ASCE, vol. 96, no. SM3, 1970)

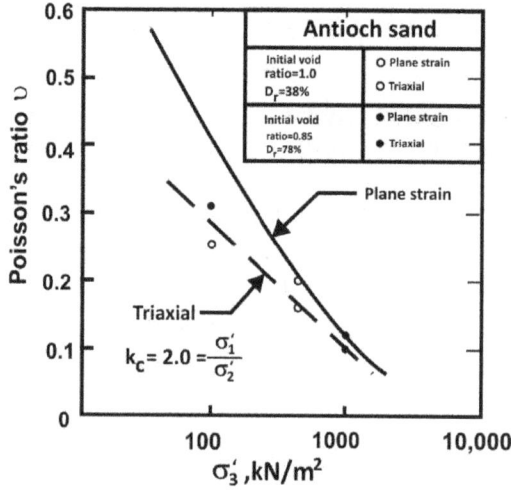

Poisson's ratio from drained tests on Antioch sand. (Redrawn after K.L. Lee, Comparison of Plane Strain and Triaxial Tests on Sand, J. Soil Mech. Found. Div., ASCE, vol. 96, no. SM3, 1970)

$$\Delta H = H\epsilon_1$$

$$\Delta B = B\epsilon_2$$

$$\Delta L = L\epsilon_3$$

*Where $H, L, B$ = height, length, and width of specimen*

*$\Delta H, \Delta B, \Delta L$ = change in height, length, and width of spicimen due to application of stresses*

*$\epsilon_1, \epsilon_2, \epsilon_3$ = strains in direction of major, intermediate and minor principal stresses*

The volume of the specimen before load application is equal to $V = LBH$, and the volume of the specimen after the load application is equal to $V - \Delta V$. Thus,

$$\Delta V = V - (V - \Delta V) = LBH - (L - \Delta L)(B - \Delta B)(H - \Delta H)$$
$$= LBH - LBH(1 - \epsilon_1)(1 - \epsilon_2)(1 - \epsilon_3)$$

Where $\Delta V$ is change in volume. Neglecting the higher order terms such as $\epsilon_1\epsilon_2, \epsilon_2\epsilon_3, \epsilon_3\epsilon_1$, and $\epsilon_1, \epsilon_2, \epsilon_3$, equation gives.

$$V = \frac{\Delta V}{V} = \epsilon_1 + \epsilon_2 + \epsilon_3$$

Where v is the change in volume per unit volume of the specimen.

For triaxial tests, $\epsilon_2 = \epsilon_3$, and they are expansions (negative sigh). So, $\epsilon_2 = \epsilon_3 = -v\epsilon_1$. Substituting this into equation above we get $V = \epsilon_1(1 - 2v)$, or

$$V = \frac{1}{2}\left(1 - \frac{v}{\epsilon_1}\right) \qquad (for\ triaxial\ test\ conditions)$$

With plane strain loading conditions, $\epsilon_2 = 0$ *and* $\epsilon_3 = -v\epsilon_1$. Hence, from equation earlier, $V = \epsilon_1(1-v)$, or

$$V = 1 - \frac{v}{\epsilon_1}\left(for\ test\ conditions\right)$$

Figure shows that for a given value of $\sigma'_3$, the Poisson's ratio obtained from plane strain loading is higher than that obtained from triaxial loading.

Hence, based on the available information at this time, it can be concluded that $\phi_p$ exceeds the value of $\phi_t$ by $0\ to\ 8°$. The greatest difference is associated with dense sands at low confining pressures. The smaller differences are associated with loose sands at all confining pressures, or dense sand at high confining pressures. Although still disputed, based on the studies described in this chapter several suggestions have been made to use a value of $\phi \approx \phi_p = 1.1\phi_t$ for calculation of the bearing capacity of strip foundations. For rectangular foundations, the stress conditions on the soil cannot be approximated by either triaxial or plane strain loadings. Meyerhof (1963) suggested for this case that the friction angle to be used for calculation of the ultimate bearing capacity should be approximated as

$$\phi = \left(1.1 - 0.1\frac{B_f}{L_f}\right)\phi_t$$

Where $L_f$ is the length of foundation and $B_f$ the width of foundation.

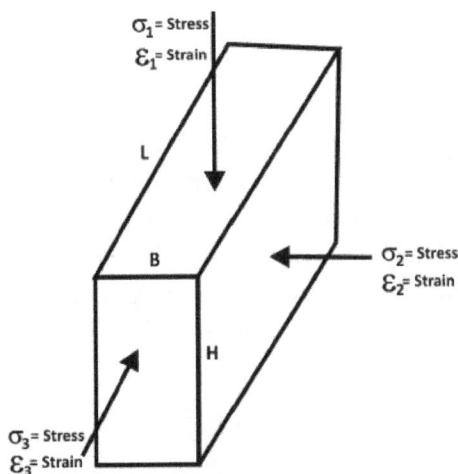

## Proctor Compaction Test

The Proctor compaction test is a laboratory method of experimentally determining the optimal moisture content at which a given soil type will become most dense and achieve its maximum dry density. The term Proctor is in honor of R. R. Proctor, who in 1933 showed that the dry density of a soil for a given compactive effort depends on the amount of water the soil contains during soil compaction. His original test is most commonly referred to as the standard Proctor compaction test; later on, his test was updated to create the modified Proctor compaction test.

These laboratory tests generally consist of compacting soil at known moisture content into a cylindrical mold of standard dimensions using a compactive effort of controlled magnitude. The soil is usually compacted into the mold to a certain amount of equal layers, each receiving a number of blows from a standard weighted hammer at a specified height. This process is then repeated for various moisture contents and the dry densities are determined for each. The graphical relationship of the dry density to moisture content is then plotted to establish the compaction curve. The maximum dry density is finally obtained from the peak point of the compaction curve and its corresponding moisture content, also known as the optimal moisture content.

The testing described is generally consistent with the American Society for Testing and Materials (ASTM) standards, and are similar to the American Association of State Highway and Transportation Officials (AASHTO) standards. Currently, the procedures and equipment details for the standard Proctor compaction test is designated by ASTM D698 and AASHTO T99. Also, the modified Proctor compaction test is designated by ASTM D1557 and AASHTO T180.

## History

Proctor's fascination with geotechnical engineering began when taking his undergraduate studies at University of California, Berkeley. He was interested in the publications of Sir Alec Skempton and his ideas on in situ behavior of natural clays. Skempton formulated concepts and porous water coefficients that are still widely used today. It was Proctor's idea to take this concept a step further and formulate his own experimental conclusions to determine a solution for the in situ behaviors of clay and ground soils that cause it to be unsuitable for construction. His idea, which was later adopted and expounded upon by Skempton, involved the compaction of the soil to establish the maximum practically-achievable density of soils and aggregates (the "practically" stresses how the value is found experimentally and not theoretically).

In the early 1930s, he finally created a solution for determining the maximum density of soils. Ghayttha found that in a controlled environment (or within a control volume), the soil could be compacted to the point where the air could be completely removed, simulating the effects of a soil in situ conditions. From this, the dry density could be determined by simply measuring the weight of the soil before and after compaction, calculating the moisture content, and furthermore calculating the dry density. Ralph R. Proctor went on to teach at the University of Arkansas.

In 1958, the modified Proctor compaction test was developed as an ASTM standard. A higher and more relevant compaction standard was necessary. There were larger and heavier compaction equipment, like large vibratory compactors and heavier steel-face rollers. This equipment could produce higher dry densities in soils along with greater stability. These improved properties allowed for the transport of far heavier truck loads over roads and highways. During the 1970s and early 1980s the modified Proctor test became more widely used as a modern replacement for the standard Proctor test.

## Theory of Soil Compaction

Compaction can be generally defined as the densification of soil by the removal of air and rearrangement of soil particles through the addition of mechanical energy. The energy exerted by compaction forces the soil to fill available voids, and the additional frictional forces between the soil particles improves the mechanical properties of the soil. Because a wide range of particles are needed in order to fill all available voids, well-graded soils tend to compact better than poorly graded soils.

The degree of compaction of a soil can be measured by its dry unit weight, $\gamma_d$. When water is added to the soil, it functions as a softening agent on the soil particles, causing them to slide between one another more easily. At first, the dry unit weight after compaction increases as the moisture content ($\omega$) increases, but after the optimum moisture content ($\omega_{opt}$) percentage is exceeded, any added water will result in a reduction in dry unit weight because the pore water pressure (pressure of water in-between each soil particle) will be pushing the soil particles apart, decreasing the friction between them.

## Comparison of Tests

The original Proctor test, ASTM D698 / AASHTO T99, uses a 4-inch-diameter (100 mm) mould which holds 1/30 cubic feet of soil, and calls for compaction of three separate lifts of soil using 25 blows by a 5.5 lb hammer falling from 12 inches, for a compactive effort of 12,400 ft-lbf/ft³. The "Modified Proctor" test, ASTM D1557 / AASHTO T180, uses same mould, but uses a 10 lb. hammer falling through 18 inches, with 25 blows on each of five lifts, for a compactive effort of about 56,250 ft-lbf/ft³. Both tests allow the use of a larger mould, 6 inches in diameter and holding 1/13.333 ft³, if the soil or aggregate contains too large a proportion of gravel-sized particles to allow repeatability with the 4-inch mould. To ensure the same compactive effort, the number of blows per lift is increased to 56.

## Alternative Compaction Testing

The California Department of Transportation has developed a similar test, California Test 216, which measures the maximum wet density, and controls the compactive effort based on the length, not the volume, of the test sample. The primary advantage of this test is that maximum density test results are available sooner, as evaporation of the compacted sample is not necessary.

## Cone Penetration Test

A CPT truck operated by the USGS.

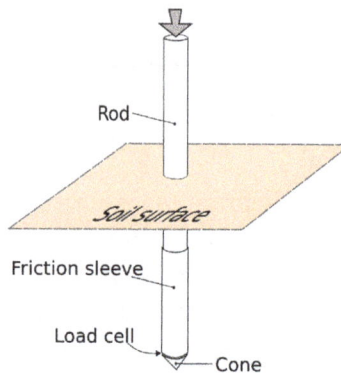
Simplified version of a cone penetrometer.

The cone penetration or cone penetrometer test (CPT) is a method used to determine the geotechnical engineering properties of soils and delineating soil stratigraphy. It was initially developed in the 1950s at the Dutch Laboratory for Soil Mechanics in Delft to investigate soft soils. Based on this history it has also been called the "Dutch cone test". Today, the CPT is one of the most used and accepted soil methods for soil investigation worldwide.

The test method consists of pushing an instrumented cone, with the tip facing down, into the ground at a controlled rate (controlled between 1.5 -2.5 cm/s accepted). The resolution of the CPT in delineating stratigraphic layers is related to the size of the cone tip, with typical cone tips having a cross-sectional area of either 10 or 15 cm$^2$, corresponding to diameters of 3.6 and 4.4 cm. A very early ultra-miniature 1 cm$^2$ subtraction penetrometer was developed and used on a US mobile ballistic missile launch system (MGM-134 Midgetman) soil/structure design program in 1984 at the Earth Technology Corporation of Long Beach, California.

## History and Development

The early applications of CPT mainly determined the soil geotechnical property of bearing capacity. The original cone penetrometers involved simple mechanical measurements of the total penetration resistance to pushing a tool with a conical tip into the soil. Different methods were employed to separate the total measured resistance into components generated by the conical tip (the "tip friction") and friction generated by the rod string. A friction sleeve was added to quantify this component of the friction and aid in determining soil cohesive strength in the 1960s. Electronic measurements began in 1948 and improved further in the early 1970s. Most modern electronic CPT cones now also employ a pressure transducer with a filter to gather pore water pressure data. The filter is usually located either on the cone tip (the so-called U1 position), immediately behind the cone tip (the most common U2 position) or behind the friction sleeve (U3 position). Pore water pressure data aids determining stratigraphy and is primarily used to correct tip friction values for those effects. CPT testing which also gathers this piezometer data is called CPTU testing. CPT and CPTU testing equipment generally advances the cone using hydraulic rams mounted on either a heavily ballasted vehicle or using screwed-in anchors as a counter-force. One advantage of CPT over the Standard Penetration Test (SPT) is a more continuous profile of soil parameters, with data recorded at intervals typically of 20 cm but as small as 1 cm.

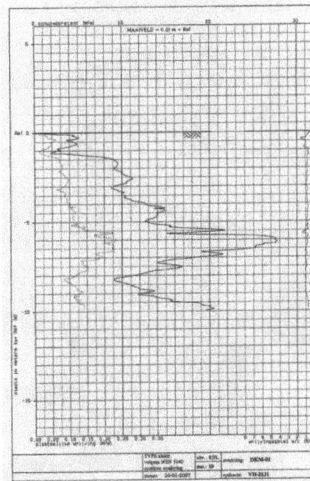

The result of a cone penetration test: resistance and friction on the left, friction ratio (%) on the right.

Manufacturers of cone penetrometer probes and data acquisition systems include Hogentogler, which has been acquired by the Vertek Division of Applied Research Associates, GeoPoint Systems BV and Pagani Geotechnical Equipment.

## Additional in Situ Testing Parameters

In addition to the mechanical and electronic cones, a variety of other CPT-deployed tools have been developed over the years to provide additional subsurface information. One common tool advanced during CPT testing is a geophone set to gather seismic shear wave and compression wave velocities. This data helps determine the shear modulus and Poisson's ratio at intervals through the soil column for soil liquefaction analysis and low-strain soil strength analysis. Engineers use the shear wave velocity and shear modulus to determine the soil's behavior under low-strain and vibratory loads. Additional tools such as laser-induced fluorescence, X-ray fluorescence, soil conductivity/resistivity, pH, temperature and membrane interface probe and cameras for capturing video imagery are also increasingly advanced in conjunction with the CPT probe.

An additional CPT deployed tool used in Britain, Netherlands, Germany, Belgium and France is a piezocone combined with a tri-axial magnetometer. This is used to attempt to ensure that tests, boreholes, and piles, do not encounter unexploded ordnance (UXB) or duds. The magnetometer in the cone detects ferrous materials of 50 kg or larger within a radius of up to about 2 m distance from the probe depending on the material, orientation and soil conditions.

## Standards and Use

CPT for geotechnical applications was standardized in 1986 by ASTM Standard D 3441 (ASTM, 2004). ISSMGE provides international standards on CPT and CPTU. Later ASTM Standards have addressed the use of CPT for various environmental site characterization and groundwater monitoring activities. For geotechnical soil investigations, CPT is more popular compared to SPT as a method of geotechnical soil investigation.Its increased accuracy, speed of deployment, more continuous soil profile and reduced cost over other soil testing methods. The ability to advance additional in situ testing tools using the CPT direct push drilling rig, including the seismic tools described above, are accelerating this process.

## Standard Penetration Test

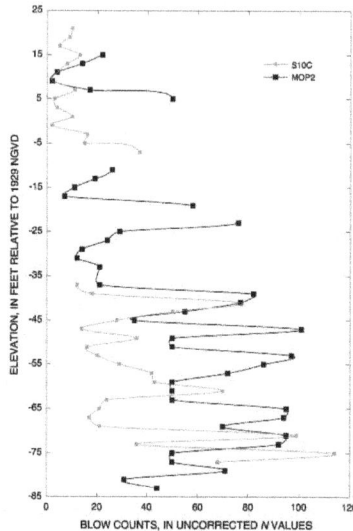

Standard penetration test N values from a surficial aquifer in south Florida.

The standard penetration test (SPT) is an in-situ dynamic penetration test designed to provide information on the geotechnical engineering properties of soil. The test procedure is described in ISO 22476-3, ASTM D1586 and Australian Standards AS 1289.6.3.1.

## Procedure

The test uses a thick-walled sample tube, with an outside diameter of 50.8 mm and an inside diameter of 35 mm, and a length of around 650 mm. This is driven into the ground at the bottom of a borehole by blows from a slide hammer with a mass of 63.5 kg (140 lb) falling through a distance of 760 mm (30 in). The sample tube is driven 150 mm into the ground and then the number of blows needed for the tube to penetrate each 150 mm (6 in) up to a depth of 450 mm (18 in) is recorded. The sum of the number of blows required for the second and third 6 in. of penetration is termed the "standard penetration resistance" or the "N-value". In cases where 50 blows are insufficient to advance it through a 150 mm (6 in) interval the penetration after 50 blows is recorded. The blow count provides an indication of the density of the ground, and it is used in many empirical geotechnical engineering formulae.

## Purpose

The main purpose of the test is to provide an indication of the relative density of granular deposits, such as sands and gravels from which it is virtually impossible to obtain undisturbed samples. The great merit of the test, and the main reason for its widespread use is that it is simple and inexpensive. The soil strength parameters which can be inferred are approximate, but may give a useful guide in ground conditions where it may not be possible to obtain borehole samples of adequate quality like gravels, sands, silts, clay containing sand or gravel and weak rock. In conditions where the quality of the undisturbed sample is suspect, e.g., very silty or very sandy clays, or hard clays, it is often advantageous to alternate the sampling with standard penetration tests to check the strength. If the samples are found to be unacceptably disturbed, it may be necessary to use a different method for measuring strength like the plate test. When the test is carried out in granular soils below groundwater level, the soil may become loosened. In certain circumstances, it can be useful to continue driving the sampler beyond the distance specified, adding further drilling rods as necessary. Although this is not a standard penetration test, and should not be regarded as such, it may at least give an indication as to whether the deposit is really as loose as the standard test may indicate.

The usefulness of SPT results depends on the soil type, with fine-grained sands giving the most useful results, with coarser sands and silty sands giving reasonably useful results, and clays and gravelly soils yielding results which may be very poorly representative of the true soil conditions. Soils in arid areas, such as the Western United States, may exhibit natural cementation. This condition will often increase the standard penetration value.

The SPT is used to provide results for empirical determination of a sand layer's susceptibility to soil liquefaction, based on research performed by Harry Seed, T. Leslie Youd, and others.

## Correlation with Soil Mechanical Properties

Despite its many flaws, it is usual practice to correlate SPT results with soil properties relevant for geotechnical engineering design. SPT results are in-situ field measurements, and not as subject to sample disturbance, and are often the only test results available, therefore the use of correlations has become common practice in many countries.

## Problems with SPT

The Standard Penetration Test recovers a highly disturbed sample, which is generally not suitable for tests which measure properties of the in-situ soil structure, such as density, strength, and consolidation characteristics. To overcome this limitation, the test is often run with a larger sampler with a slightly different tip shape, so the disturbance of the sample is minimized, and testing of structural properties is meaningful for all but soft soils. However, this results in blow counts which are not easily converted to SPT N-values – many conversions have been proposed, some of which depend on the type of soil sampled, making reliance on blow counts with non-standard samplers problematic.

Standard Penetration Test blow counts do not represent a simple physical property of the soil, and thus must be correlated to soil properties of interest, such as strength or density. There exist multiple correlations, none of which are of very high quality. Use of SPT data for direct prediction of liquefaction potential suffers from roughness of correlations and from the need to "normalize" SPT data to account for overburden pressure, sampling technique, and other factors. Additionally, the method cannot collect accurate data for weak soil layers for several reasons:

1.  The results are limited to whole numbers for a specific driving interval, but with very low blow counts, the granularity of the results, and the possibility of a zero result, makes handling the data cumbersome.

2.  In loose sands and very soft clays, the act of driving the sampler will significantly disturb the soil, including by soil liquefaction of loose sands, giving results based on the disturbed soil properties rather than the intact soil properties.

A variety of techniques have been proposed to compensate for the deficiencies of the standard penetration test, including the Cone penetration test, in-situ vane shear tests, and shear wave velocity measurements.

## References

*   Myers, Ken S. (2008), "Wikimmunity: Fitting the Communications Decency Act to Wikipedia", Harvard Journal of Law and Technology, The Berkman Center for Internet and Society, 20: 163, SSRN 916529

- Head, K.H. (1998). Effective Stress Tests, Volume 3, Manual of Soil Laboratory Testing, (2nd ed.). John Wiley & Sons. ISBN 978-0-471-97795-7

- Cubric, Marija (2007). "Analysis of the use of Wiki-based collaborations in enhancing student learning". University of Hertfordshire. Retrieved April 25, 2011

- Lombardo, Nancy T. (June 2008). "Putting Wikis to Work in Libraries". Medical Reference Services Quarterly. 27 (2): 129–145. doi:10.1080/02763860802114223. Archived from the original on November 29, 2012

- Reddy, K.R.; Saxena, S.K.; Budiman, J.S. (June 1992). "Development of A True Triaxial Testing Apparatus" (pdf). Geotechnical Testing Journal. ASTM. 15 (2): 89–105

- BS 1377-1 (1990). Methods of test for soils for civil engineering purposes. General requirements and sample preparation. BSI. ISBN 0-580-17692-4

- Vardoulakis, I. (1979). "Bifurcation analysis of the triaxial test on sand samples". Acta Mechanica. 32: 35. doi:10.1007/BF01176132

- De Reister, J., 1971, "Electric Penetrometer for Site Investigations"; Journal of SMFE Division, ASCE, Vol. 97, SM-2, pp. 457-472

- Bill Venners (October 20, 2003). "Exploring with Wiki: A Conversation with Ward Cunningham, Part I". artima developer. Retrieved December 12, 2014

- Ibsen, L.B. (1994). "The stable state in cyclic triaxial testing on sand". Soil Dynamics and Earthquake Engineering. 13: 63. doi:10.1016/0267-7261(94)90042-6

- Goldman, Eric, "Wikipedia's Labor Squeeze and its Consequences", Journal on Telecommunications and High Technology Law, 8

- Price, D.G. (2009). De Freitas, M.H., ed. Engineering Geology: Principles and Practice. Springer. p. 450. ISBN 3-540-29249-7

- ASTM D4767-11 (2011). Standard Test Method for Consolidated Undrained Triaxial Compression Test for Cohesive Soils. ASTM International, West Conshohocken, PA, 2003. doi:10.1520/D4767-11

# Fundamentals of Soil Consolidation

The process by which the volume of soil decreases is known as is called soil consolidation. An oedometer test can analyze the compression of soil. One-dimensional consolidation and secondary consolidation have also been mentioned. The aspects elucidated in this chapter are of vital importance, and provide a better understanding of geotechnical engineering and soil science.

## Consolidation (Soil)

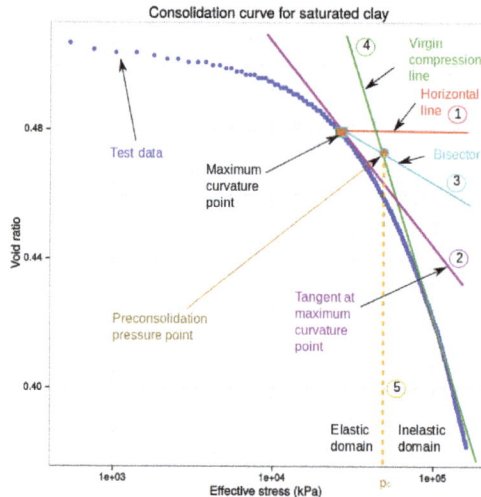

Consolidation curve for saturated clay

The experimentally determined consolidation curve (blue dots) for a saturated clay showing a procedure for computing the preconsolidation stress.

Consolidation is a process by which soils decrease in volume. According to Karl von Terzaghi "consolidation is any process which involves a decrease in water content of saturated soil without replacement of water by air." In general it is the process in which reduction in volume takes place by expulsion of water under long term static loads. It occurs when stress is applied to a soil that causes the soil particles to pack together more tightly, therefore reducing its bulk volume. When this occurs in a soil that is saturated with water, water will be squeezed out of the soil. The magnitude of consolidation can be predicted by many different methods. In the Classical Method, developed by Terzaghi, soils are tested with an oedometer test to determine their compression index. This can be used to predict the amount of consolidation.

When stress is removed from a consolidated soil, the soil will rebound, regaining some of the volume it had lost in the consolidation process. If the stress is reapplied, the soil will consolidate again along a recompression curve, defined by the recompression index. The soil which had its load removed is considered to be *overconsolidated*. This is the case for soils which have previously had glaciers on them. The highest stress that it has been subjected to is termed the *preconsolidation stress*. The *over consolidation ratio* or OCR is defined as the highest stress experienced divided by the current stress. A soil which is currently experiencing its highest stress is said to be *normally consolidated* and to have an OCR of one. A soil could be considered *underconsolidated* immediately after a new load is applied but before the excess pore water pressure has had time to dissipate.

## Consolidation Analysis

## Spring Analogy

The process of consolidation is often explained with an idealized system composed of a spring, a container with a hole in its cover, and water. In this system, the spring represents the compressibility or the structure of the soil itself, and the water which fills the container represents the pore water in the soil.

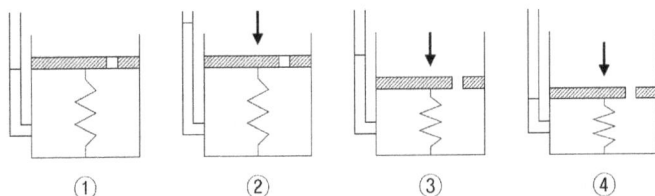

1. The container is completely filled with water, and the hole is closed. (Fully saturated soil)

2. A load is applied onto the cover, while the hole is still unopened. At this stage, only the water resists the applied load. (Development of excess pore water pressure)

3. As soon as the hole is opened, water starts to drain out through the hole and the spring shortens. (Drainage of excess pore water pressure)

4. After some time, the drainage of water no longer occurs. Now, the spring alone resists the applied load. (Full dissipation of excess pore water pressure. End of consolidation)

## Primary Consolidation

This method assumes consolidation occurs in only one-dimension. Laboratory data is used to construct a plot of strain or void ratio versus effective stress where the effective stress axis is on a logarithmic scale. The plot's slope is the compression index or recompression index. The equation for consolidation settlement of a normally consolidated soil can then be determined to be:

$$\delta_c = \frac{C_c}{1+e_0} H \log\left(\frac{\sigma'_{zf}}{\sigma'_{z0}}\right)$$

where

$\delta_c$ is the settlement due to consolidation.

$C_c$ is the compression index.

$e_0$ is the initial void ratio.

H is the height of the compressible soil.

$\sigma_{zf}$ is the final vertical stress.

$\sigma_{z0}$ is the initial vertical stress.

$C_c$ can be replaced by $C_r$ (the recompression index) for use in overconsolidated soils where the final effective stress is less than the preconsolidation stress. When the final effective stress is greater than the preconsolidation stress, the two equations must be used in combination to model both the recompression portion and the virgin compression portion of the consolidation processes, as follows,

$$\delta_c = \frac{C_r}{1+e_0} H \log\left(\frac{\sigma'_{zc}}{\sigma'_{z0}}\right) + \frac{C_c}{1+e_0} H \log\left(\frac{\sigma'_{zf}}{\sigma'_{zc}}\right)$$

where $\sigma_{zc}$ is the preconsolidation stress of the soil.

## Secondary Compression

Secondary compression is the compression of soil that takes place after primary consolidation. Even after the reduction of hydrostatic pressure some compression of soil takes place at slow rate. This is known as secondary compression. Secondary compression is caused by creep, viscous behavior of the clay-water system, compression of organic matter, and other processes. In sand, settlement caused by secondary compression is negligible, but in peat, it is very significant. Due to secondary compression some of the highly viscous water between the points of contact is forced out.

Secondary compression is given by the formula

$$S_s = \frac{H_0}{1+e_0} C_a \log\left(\frac{t}{t_{90}}\right)$$

Where $H_0$ is the height of the consolidating medium

$e_0$ is the initial void ratio

$C_a$ is the secondary compression index

t is the length of time after consolidation considered

$t_{90}$ is the length of time for achieving 90% consolidation

## Time Dependency

The time for consolidation to occur can be predicted. Sometimes consolidation can take years. This is especially true in saturated clays because their hydraulic conductivity is extremely low, and this causes the water to take an exceptionally long time to drain out of the soil. While drainage is occurring, the pore water pressure is greater than normal because it is carrying part of the applied stress (as opposed to the soil particles).

$$T_v = \frac{c_v * t}{(H_{dr})^2}$$

Where $T_v$ is the time factor.

$H_{dr}$ is the average longest drain path during consolidation.

t is the time at measurement

$C_v$ is defined as the compression index found using the log method with

$$C_v = \frac{T_{50}H_{dr}^2}{t_{50}}$$

or the root method with

$$C_v = \frac{T_{90}H_{dr}^2}{t_{90}}$$

$t_{50}$ time to 50% deformation (consolidation) and $t_{90}$ is 90%

Whereas $T_{90}$=0.848 $T_{50}$=0.197

## Oedometer Test

An oedometer test is a kind of geotechnical investigation performed in geotechnical engineering that measures a soil's consolidation properties. Oedometer tests are performed by applying different loads to a soil sample and measuring the deformation response. The results from these tests are used to predict how a soil in the field will deform in response to a change in effective stress.

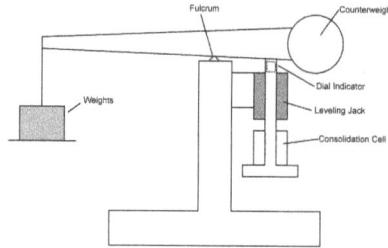

Incremental Loading frame developed by Bishop

## Concept

Oedometer tests are designed to simulate the one-dimensional deformation and drainage conditions that soils experience in the field. To simulate these conditions, rigid confining rings are used to prevent lateral displacement of the soil sample. Porous stones are placed on the top and bottom of the sample to allow drainage in the vertical direction. To better simulate one-dimensional strain, a diameter-to-height ratio in the sample of 3:1 or more is used. Because the process of consolidation involves movement of water out of a soil, it is important to prevent drying of the soil.

## History

Consolidation experiments were first carried out in 1910 by Frontard. A thin sample (2in thick by 14in in diameter) was cut and placed in a metal container with a perforated base. This sample was then loaded through a piston incrementally, allowing equilibrium to be reached after each increment. To prevent drying of the clay, the test was done in a room with high humidity.

Karl von Terzaghi started his consolidation research in 1919 at Robert College in Istanbul. Through these experiments, Terzaghi started to develop his theory of consolidation which was eventually published in 1923.

The Massachusetts Institute of Technology played a key role in early consolidation research. Both Terzaghi and Arthur Casagrande spent time at M.I.T. - Terzaghi from 1925 to 1929 and Casagrande from 1926 to 1932. During that time, the testing methods and apparatuses for consolidation testing were improved. Research was continued at MIT in the 1940s by Donald Taylor.

## Testing Procedures

There are many oedometer tests that are used to measure consolidation properties. The most common type is the Incremental Loading (IL) test.

## Incremental Loading

The text *Soil Mechanics in Engineering Practice* describes a general procedure for the

Incremental Loading test. A stiff confining ring with a sharp edge is used to cut a sample of soil directly from a larger block of soil. Excess soil is carefully carved away, leaving a sample with a diameter-to-height ratio of 3 or more. Porous stones are placed on the top and bottom of the sample to provide drainage. A rigid loading cap is then placed on top of the upper porous stone. This assembly is then placed into a loading frame.

Weights are placed on the frame, imposing a load on the soil. Compression of the sample is measured over time by a dial indicator. By observing the deflection value over time data, it can be determined when the sample has reached the end of primary consolidation. Another load is then immediately placed on the soil and this process is repeated. After a significant total load has been applied, the load on the sample is decreased incrementally. Using a load increment ratio of 1/2 provides a sufficient number of data points to describe the relationship between void ratio and effective stress for a soil.

ASTM International has a standard testing procedure for incremental loading: ASTM D2435-04. Other testing standards such as BS 1377:5, ASTM D3877, ASTM D4546 and AASHTO T216 provide procedures for conducting tests for determination of the consolidation characteristics of soils. Oedometer test set is required to perform the test. It is used to determine the consolidation characteristics of soils of low permeability. Tests are carried out on specimens prepared from undisturbed samples. Ideally, following would be needed to perform the Oedometer test:

- 1 x Bench

- 3 x Oedometers

- 3 x Cells, either 50mm or 63.5mm, or 75mm

- 3 x Dial gauges, either analogue, or digital

- 1 x Weight set

# Preconsolidation Pressure

Preconsolidation pressure is the maximum effective vertical overburden stress that a particular soil sample has sustained in the past. This quantity is important in geotechnical engineering, particularly for finding the expected settlement of foundations and embankments. Alternative names for the preconsolidation pressure are preconsolidation stress, pre-compression stress, pre-compaction stress, and preload stress. A soil is called overconsolidated if the current effective stress acting on the soil is less than the historical maximum.

The preconsolidation pressure can help determine the largest overburden pressure that can be exerted on a soil without irrecoverable volume change. This type of volume change is important for understanding shrinkage behavior, crack and structure formation and resistance to shearing stresses. Previous stresses and other changes in a soil's history are preserved within the soil's structure. If a soil is loaded beyond this point the soil is unable to sustain the increased load and the structure will break down. This breakdown can cause a number of different things depending on the type of soil and its geologic history.

Preconsolidation pressure cannot be measured directly, but can be estimated using a number of different strategies. Samples taken from the field are subjected to a variety of tests, like the constant rate of strain test (CRS) or the incremental loading test (IL). These tests can be costly due to expensive equipment and the long period of time they require. Each sample must be undisturbed and can only undergo one test with satisfactory results. It is important to execute these tests precisely to ensure an accurate resulting plot. There are various methods for determining the preconsolidation pressure from lab data. The data is usually arranged on a semilog plot of the effective stress (frequently represented as $\sigma'vc$) versus the void ratio. This graph is commonly called the e log p curve or the consolidation curve.

## Methods

The preconsolidation pressure can be estimated in a number of different ways but not measured directly. It is useful to know the range of expected values depending on the type of soil being analyzed. For example, in samples with natural moisture content at the liquid limit (liquidity index of 1), preconsolidation ranges between about 0.1 and 0.8 tsf, depending on soil sensitivity (defined as the ratio of undisturbed peak undrained shear strength to totally remolded undrained shear strength). For natural moisture at the plastic limit (liquidity index equal to zero), preconsolidation ranges from about 12 to 25 tsf.

## Arthur Casagrande's Graphical Method

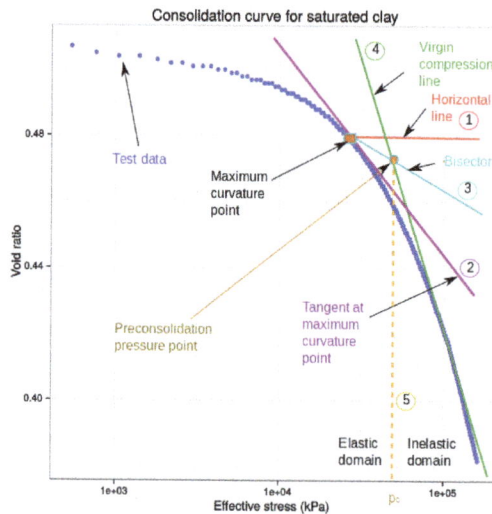

The consolidation curve for a saturated clay showing
the procedure for finding the preconsolidation pressure.

Using a consolidation curve:(Casagrande 1936)

1.  Choose by eye the point of maximum curvature on the consolidation curve.

2.  Draw a horizontal line from this point.

3.  Draw a line tangent to the curve at the point found in part 1.

4.  Bisect the angle made from the horizontal line in part 2 and the tangent line in part 3.

5.  Extend the "straight portion" of the virgin compression curve (high effective stress, low void ratio: almost vertical on the right of the graph) up to the bisector line in part 4.

The point where the lines in part 4 and part 5 intersect is the preconsolidation pressure.

Gregory et al. proposed an analytical method to calculate preconsolidation stress that avoids subjective interpretations of the location of the maximum curvature point (i.e. Minimum radius of curvature). Tomás et al. used this method to calculate the preconsolidation pressure of 139 undisturbed soil samples to generate preconsolidation pressure maps of the Vega Baja of the Segura (Spain).

## Estimation of the "Most Probable" Preconsolidation Pressure

Using a consolidation curve, intersect the horizontal portion of the recompression curve and a line tangent to the compression curve. This point is within the range of probable preconsolidation pressures. It can be used in calculations that require less accuracy or if a rough estimate is all that is required.

## Mechanisms Causing Preconsolidation

Various different factors can cause a soil to approach its preconsolidation pressure:

*   Change in total stress due to removal of overburden can cause preconsolidation pressure in a soil. For example, removal of structures or glaciation would cause a change in total stress that would have this effect.

*   Change in pore water pressure: A change in water table elevation, Artesian pressures, deep pumping or flow into tunnels, and desiccation due to surface drying or plant life can bring soil to its preconsolidation pressure.

*   Change in soil structure due to aging (secondary compression): Over time, soil will consolidate even after high pressures from loading and pore water pressure have been depleted.

*   Environmental changes: Changes in pH, temperature, and salt concentration can cause a soil to approach its preconsolidation pressure.

- Chemical weathering: Different types of chemical weathering will cause pre-consolidation pressure. Precipitation, cementing agents, and ion exchange are a few examples.

## Uses

Preconsolidation pressure is used in many calculations of soil properties essential for structural analysis and soil mechanics. One of the primary uses is to predict settlement of a structure after loading. This is required for any construction project such as new buildings, bridges, large roads and railroad tracks. All of these require site evaluation before construction. Preparing a site for construction requires an initial compression of the soil to prepare for foundation to be added. It is important to know the preconsolidation pressure because it will help to determine the amount of loading that is appropriate for the site. It will also help to determine whether recompression (after excavation), if the conditions allow, soil can exhibit volumetric expansion, recompression, due to the removal of load conditions need to be considered.

# One-dimensional Consolidation

## General Concepts of One-dimensional Consolidation

To understand the basic concepts of consolidation, consider a clay layer of thickness $H_t$ located below the groundwater level and between two highly permeable sand layers as shown in Figure. If a surcharge of intensity $\Delta\sigma$ is applied at the ground surface over a very large area, the pore water pressure in the clay layer will increase. For a surcharge of infinite extent, the immediate increase of the pore water pressure, $\Delta u$, at all depths of the clay layer will be equal to the increase of the total stress, $\Delta\sigma$. Thus, immediately after the application of the surcharge.

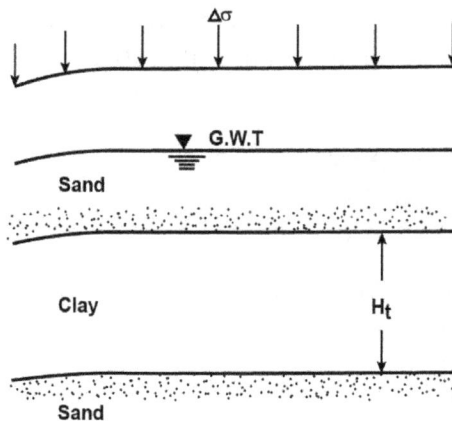

$$\Delta u = \Delta \sigma$$

Since the total stress is equal to the sum of the effective stress and the pore water pressure at all depth soft the clay layer the increase of effective stress due to the surcharge (immediately after application) will be equal to zero (i.e., $\Delta\sigma' = 0$ where $\Delta\sigma'$ is the increase of the effective stress). In other words, at time t = 0, the entire stress increase at all depths of the clay is taken by the pore water pressure and none b y the soil skeleton. This is shown in Figure a. (It must be pointed out that, for loads applied over a limited area, it may to be true that the increase of the pore water pressure is equal to the increase of vertical stress at any depth at time t = 0.

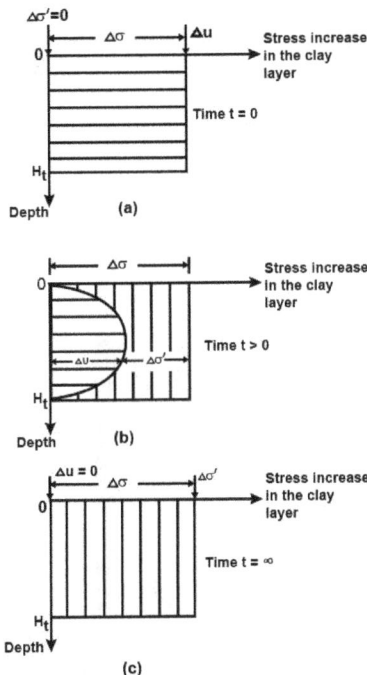

Change of pore water pressure and effective
stress in the clay layer shown in Figure due to the surcharge

After application of the surcharge (i.e., at time $t > 0$ ), the water in the void spaces of the clay layer will be squeezed out and will flow toward both the highly permeable sand layers, thereby reducing the excess pore water pressure. This, in turn, will increase the effective stress by an amount since $\Delta\sigma' + \Delta u = \Delta\sigma$. Thus, at time $t > 0$,

$$\Delta\sigma' > 0$$

And $\Delta u < \Delta\sigma$

This fact is shown in Figure b.

Theoretically, at time $t = \infty$ the excess pore water pressure at all depths of the clay layer will be dissipated by gradual drainage. Thus, at time $t = \infty$,

$$\Delta\sigma' = \Delta\sigma$$

And $\Delta u = 0$

This shown in Figure c.

This gradual process of increase of effective stress in the clay layer due to the surcharge will result in a settlement which is time-dependent and is referred to as the process of consolidation.

## Theory of One-Dimensional Consolidation

The theory for the time rate of one-dimensional consolidation was first proposed by Terzaghi (1925). The underlying assumption in the derivation of the mathematical equations are as follows:

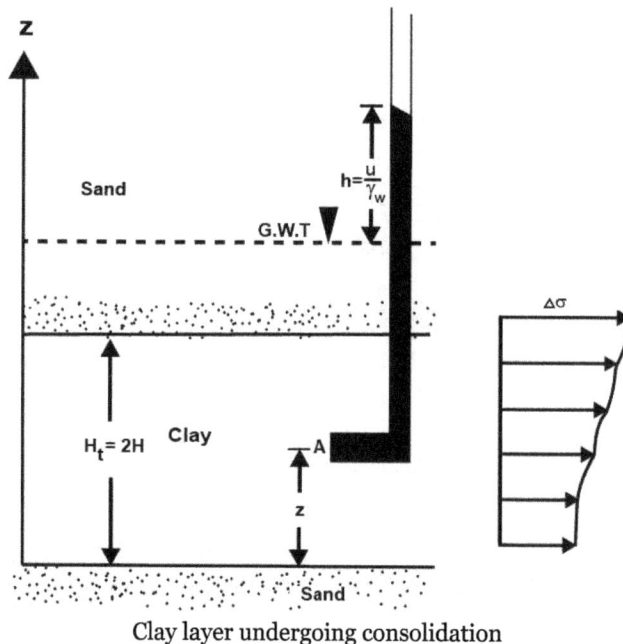

Clay layer undergoing consolidation

1. The clay layer is homogeneous.

2. The clay layer is saturated.

3. The compression of the soil layer is due to the change in volume only, which, in turn, is due to the squeezing out of water from the void spaces.

4. Darcy's law valid.

5. Deformation of soil occurs only in the direction of the load application.

6. The coefficient of consolidation $C_v$ is constant during the consolidation.

With the above assumptions, let us consider a clay layer of thickness $H_1$ as shown in Figure. The layer is located between two highly permeable sand layers. In this case of one-dimensional consolidation, the flow of water into and out of the soil element is in one direction only, i.e., in the z direction. This means that $q_x$, $q_y$, $dq_x$ and $dq_y$ are equal to zero, and thus the rate of low into and out of the soil element can be given by:

$(q_z + dq_z) - q_z$ = rate of change of volume of soil element

$$= \frac{\partial v}{\partial t} \tag{1}$$

Where $V = dx\, dy\, dz$ $\tag{2}$

we obtain

$$k\frac{\partial^2 h}{\partial z^2} dx\, dy\, dz = \frac{\partial v}{\partial t} \tag{3}$$

Where k is the coefficient of permeability [$k=k_z$]. However,

$$h = \frac{u}{\gamma_w} \tag{4}$$

where $\gamma_w$ is the unit weight of water. Substitution of equation (4) and (3) and rearranging gives

$$\frac{k}{\gamma_w}\frac{\partial^2 u}{\partial z^2} = \frac{1}{dx\, dy\, dz}\frac{\partial V}{\partial t} \tag{5}$$

During consolidation the rate of change of volume is equal to the rate of change of the void volume. So,

$$\frac{\partial v}{\partial t} = \frac{\partial V_v}{\partial t} \tag{6}$$

Where $V_v$ is the volume of voids in the soil element. But

$$V_V = eV_s \tag{7}$$

Where $V_s$ is the volume of soil solids in the element, which is constant, and e is the void ratio. So,

$$\frac{\partial V}{\partial t} = V_s\frac{\partial e}{\partial t} = \frac{v}{1+e}\frac{\partial e}{\partial t} = \frac{dx\, dy\, dz}{1+e}\frac{\partial e}{\partial t} \tag{8}$$

Substituting the above relation into equation (5), we get

$$\frac{k}{\gamma_w}\frac{\partial^2 u}{\partial z^2} = \frac{1}{1+e}\frac{\partial e}{\partial t} \tag{9}$$

The change in void ratio, $\partial e$, is due to the increase of effective stress; assuming that these are linearly related, then

$$\partial e = -a_v \partial(\Delta\sigma') \tag{10}$$

Combining equations (9) and (10),

$$\frac{k}{\gamma_w}\frac{\partial^2 u}{\partial z^2} = \frac{a_v}{1+e}\frac{\partial u}{\partial t} = m_v\frac{\partial u}{\partial t} \tag{11}$$

Where $m_v$ =coefficient of volume compressibility = $\dfrac{\partial_v}{1+e}$ (12)

Or $\dfrac{\partial u}{\partial t} = \dfrac{k}{\gamma_w m_v \partial z_2} = C_v\dfrac{\partial^2 u}{\partial z^2}$ (13)

Where $C_v$ =coefficient of consolidation = $\dfrac{k}{\gamma_w m_v}$ (14)

Equation (13) is the basic differential equation of Terzaghi's consolidation theory and can be solved with proper boundary conditions. To solve the equation, assume u to be the product of two functions, i.e., the product of a function of z and a function of t, or

$$u = F(z)G(t) \tag{15}$$

So, $\dfrac{\partial u}{\partial t} = F(z)\dfrac{\partial}{\partial t}G(t) = F(z)G'(t)$ (16)

And $\dfrac{\partial^2 u}{\partial z^2} = \dfrac{\partial^2}{\partial z^2}F(z)G(t) = F''(z)G(t)$ (17)

From equations (13), (16), and (17),

$F(z)G'(t) = C_v F''(z)G(t)$ or

$$\frac{F''(z)}{F(z)} = \frac{G'(t)}{C_v G(t)} \tag{18}$$

The right-hand side of equation (18) is a function of z only and is independent of t; the left-hand side of the equation is a function of t only and is independent of z. therefore, they must be equal to a constant, say- $B^2$. So,

$$F''(z) = -B^2 F(z) \tag{19}$$

A solution to equation (19) can be given by

$$F(z) = A_1 \cos Bz + A_2 \sin Bz \qquad (20)$$

Where $A_1$ and $A_2$ are constants.

Again, the right-hand side of equation (18) may be written as

$$G'(t) = -B^2 C_v G(t) \qquad (21)$$

The solution to equation (21) is given by

$$G(t) = A_3 \exp(-B^2 C_v t) \qquad (22)$$

Where $A_3$ is a constant. Combining equations (15), (20), and (22),

$$\begin{aligned} u &= (A_1 \cos Bz + A_2 \sin Bz) A_3 \exp(-B^2 C_v t) \\ &= (A_4 \cos Bz + A_5 \sin Bz) \exp(-B^2 C_v t) \end{aligned} \qquad (23)$$

Where $A_4 = A_1 A_3$ and $A_5 = A_2 A_3$.

The constants in equation (23) can be evaluated from the boundary conditions, which are as follows:

1.  At time $t = 0, u = u_t$ (initial excess pore water pressure at any depth).

2.  $u = 0$ at $z = 0$

3.  $u = 0$ at $z = H_t = 2H$.

Note that H is the length of the longest drainage path. In this case, which is two-way drainage condition (top and bottom of the clay layer), H is equal to half the total thickness of the clay layer, $H_t$.

The second boundary condition dictates that $A_4 = 0$, and from the third boundary condition we get

$$A_5 \sin 2BH = 0 \text{ or } 2BH = n\pi$$

Where n is an integer. From the above, a general solution of equation (23) can be in given the form

$$u = \sum_{n=1}^{n=\infty} A_n \sin\frac{n\pi z}{2H} \exp\left(\frac{-n^2\pi^2 T_v}{4}\right) \qquad (24)$$

Where $T_v$ is the nondimensional time factor and is equal to $C_v t / H^2$

To satisfy the first boundary condition, we must have the coefficients of $A_n$ such that

$$u = \sum_{n=1}^{n=\infty} A_n \sin \frac{n\pi z}{2H} \qquad (25)$$

Equation (25) is a Fourier sine series, and An can be given by

$$A_n = \frac{1}{H} \int_0^{2H} u_t \sin \frac{n\pi z}{2H} dz \qquad (26)$$

Combining equations (24) and (26),

$$u = \sum_{n=1}^{n=\infty} \left( \frac{1}{H} \int_0^{2H} u_t \sin \frac{n\pi z}{2H} dz \right) \sin \frac{n\pi z}{2H} \exp\left( \frac{-n^2\pi^2 T_v}{4} \right) \qquad (27)$$

So far we have not made any assumptions regarding the variation of $u_t$ with the depth of the clay layer. Several possible types of variation for $u_t$ are considered below.

Constant $u_t$ with depth. if $u_t$ is constant with depth – i.e., if $u_t = u_0$ – referring to equation (27),

$$\frac{1}{H} \int_0^{2H} \overset{u_t}{\underset{=u_0}{\uparrow}} \sin \frac{n\pi z}{2H} dz = \frac{2u_0}{n\pi}(1 - \cos n\pi)$$

So, $u = \sum_{n=1}^{n=\infty} \frac{2u_0}{n\pi}(1 - \cos n\pi) \sin \frac{n\pi z}{2H} \exp\left( \frac{-n^2\pi^2 T_v}{4} \right) \qquad (28)$

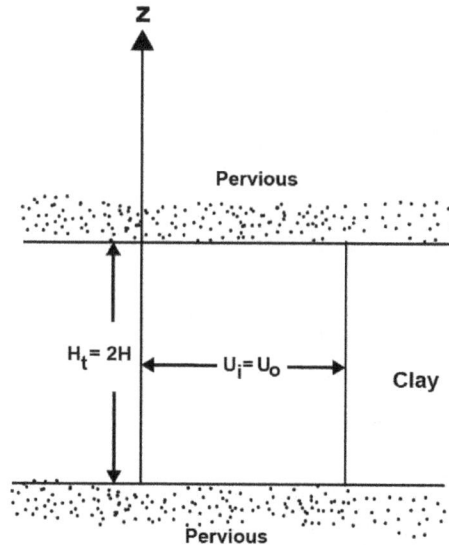

Initial excess pore water pressure-constant with depth (double drainage)

Note that the term $1 - \cos n\pi$ in the above equation is zero for cases when n is even;

therefore, u is also zero. For the nonzero terms, it is convenient to substitute n = 2m+1 where m is an integer. So equation (28) will no read

$$u = \sum_{m=0}^{m=\infty} \frac{2u_0}{(2m+1)}\left[1-\cos(2m+1)\pi\right]\sin\frac{(2m+1)\pi z}{2H} \times \exp\left[\frac{-2(m+1)^2\pi^2 T_v}{4}\right]$$

OR $u = \sum_{m=0}^{m=\infty} \frac{2u_0}{M}\sin\frac{Mz}{H}\exp(-M^2 T_v)$                   (29)

Where $M = (2m+1)\pi/2$. At a given time, the degree of consolidation at any depth z is defined as

$$U_Z = \frac{\text{excess pore water pressure dissipated}}{\text{initial excess pore water pressure}}$$

$$= \frac{u_1 - u}{u_t} = 1 - \frac{u}{u_t} = \frac{\sigma'}{u_t} = \frac{\sigma'}{u_0}$$                   (30)

Where $\sigma'$ is the increase of effective stress at a depth z due to consolidation. From equations (29) and (30),

$$U_Z = 1 - \sum_{m=0}^{m=\infty} \frac{2}{M}\sin\frac{Mz}{H}\exp(-M^2 T_v)$$                   (31)

Shows the variation of $U_Z$ with depth for various values of the non-dimensional time factor, $T_v$; these curves are called isocrones.

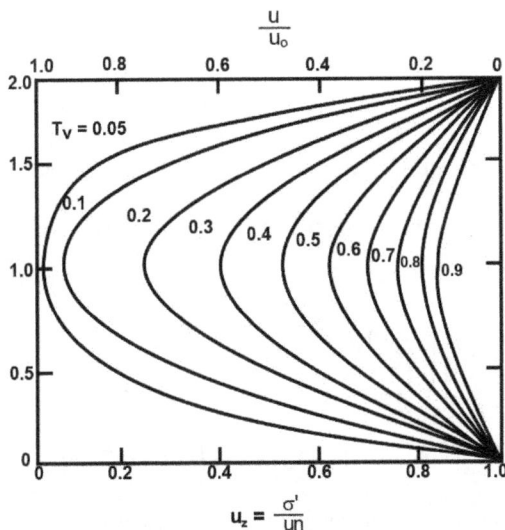

Variation of $U_Z$ with $z/H$ and $T_v$

In most cases, however, we need to obtain the average degree of consolidation for the entire layer. This is given by

$$U_{av} = \frac{(1/H_t)\int_0^{H_t} u_t dz - (1/H_t)\int_0^{H_t} u\, dz}{(1/H_t)\int_0^{H_t} u_t dz} \tag{32}$$

The average degree of consolidation is also the ratio of consolidation settlement at any time to maximum consolidation settlement. Note, in this case, that $H_t = 2H$ and $u_t = u_0$.

Combining equations (29) and (32),

$$U_{av} = 1 - \sum_{m=0}^{m=\infty} \frac{2}{M^2} \exp(-M^2 T_v) \tag{33}$$

Figure gives the variation of $U_{av}$ vs. $T_v$

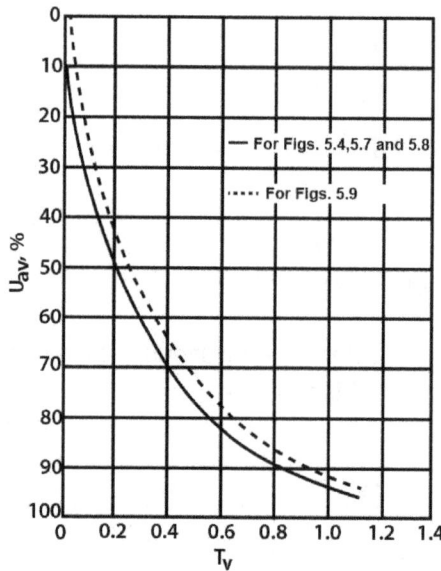

Variation of average degree of consolidation

Terzaghi suggested the following equations for $U_{av}$ to approximate the values obtained from equation (33):

For $U_{av} = 0\,to\,53\%$:   $T_v = \frac{\pi}{4}\left(\frac{U\%}{100}\right)^2 \tag{34}$

For $U_{av} = 53\,to\,100\%$:   $T_v = 1.78 - 0.933\left[\log(100 - U\%)\right] \tag{35}$

Sivaram and Swamee (1977) gave the following equation for $U_{av}$ varying from 0 to 100%:

$$\frac{U_{av}\%}{100} = \frac{(4T_v/\pi)^{0.5}}{\left[1+(4T_v/\pi)^{2.8}\right]^{0.179}} \tag{36}$$

Or $T_v = \dfrac{(\pi/4)(U_{av}\%/100)^2}{\left[1-(U_{av}\%/100)^{5.6}\right]^{0.357}}$ (37)

Equations (36) and (37) give an error in $T_v$ of less than 1% for $0\% < U_{av} < 90\%$ and less than 3% for $90\% < U_{av} < 100\%$.

Table: Variation of $T_v$ with $U_{av}$ [equation (33)]

| $U_{av},\%$ | $T_v$ | $U_{av},\%$ | $T_v$ |
|---|---|---|---|
| 0 | 0 | 60 | 0.287 |
| 10 | 0.008 | 65 | 0.342 |
| 20 | 0.031 | 70 | 0.403 |
| 30 | 0.071 | 75 | 0.478 |
| 35 | 0.096 | 80 | 0.567 |
| 40 | 0.126 | 85 | 0.684 |
| 45 | 0.159 | 90 | 0.848 |
| 50 | 0.197 | 95 | 1.127 |
| 55 | 0.238 | 100 | $\infty$ |

It must be pointed out that, if we have a situation of one-way drainage as shown in Figure a and b, equation (33) would still be valid. Note, however, that the length of the drainage path is equal to the total thickness of the clay layer.

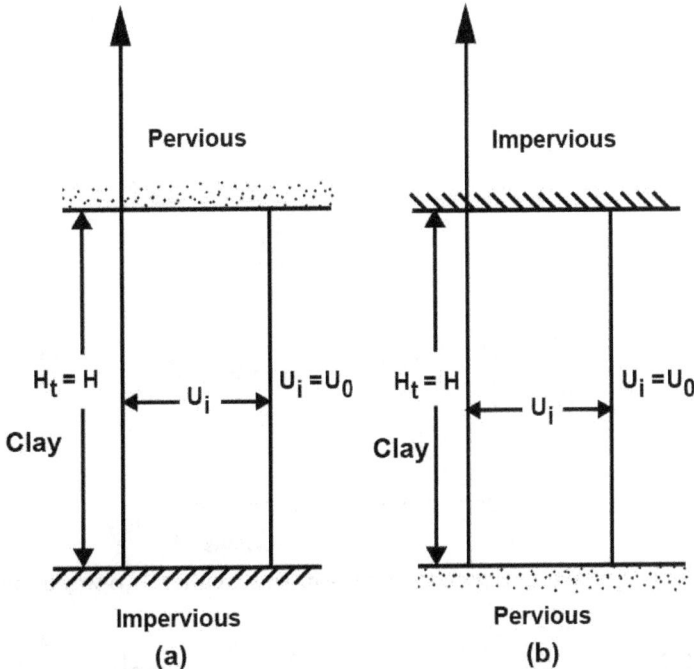

Initial excess pore pressure distribution-one way drainage, $u_i$ constant with depth

Linear variation of $u_t$. The linear variation of the initial excess pore water pressure, as shown in Figure, may be written as

$$u_t = u_1 - u_2 \frac{H - z}{H} \tag{38}$$

Substitution of the above relation for $u_t$ into equation (27) yields

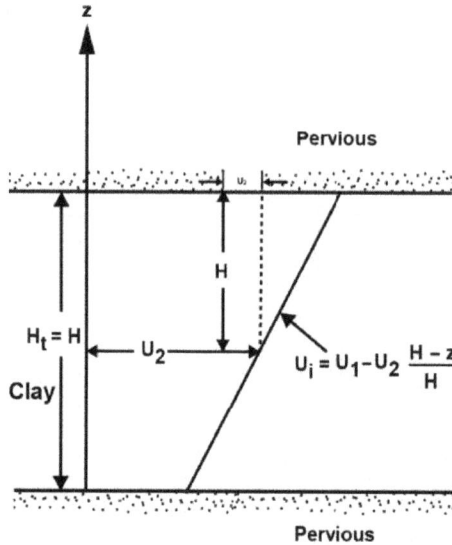

linearly varying initial excess pore water pressure distribution-two-way drainage

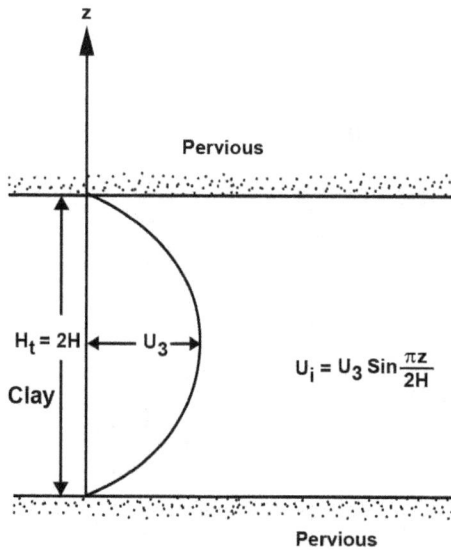

Sinusoidal initial excess pore water pressure distribution-two-way drainage

$$u = \sum_{n=1}^{n=\infty} \left[ \frac{1}{H} \int_0^{2H} \left( u_1 - u_2 \frac{H - z}{H} \right) \sin \frac{n\pi z}{2H} dz \right] \sin \frac{n\pi z}{2H} \times \exp \left( \frac{-n^2 \pi^2 T_v}{4} \right) \tag{39}$$

The average degree of consolidation can be obtained by solving equations (39) and (32):

$$U_{av} = 1 - \sum_{m=0}^{m=\infty} \frac{2}{M^2} \exp(-M^2 T_v)$$

This is identical to equation (33), which was for the case where the excess pore water pressure is constant with depth, and so the same curves as given in Figure can be used.

Sinusoidal variation of $u_t$. Sinusoidal variation can be represented by the equation

$$u_t = u_3 \sin \frac{\pi z}{2H} \qquad (40)$$

The solution for the average degree of consolidation for this type of excess pore water pressure distribution is of the form

$$U_{av} = 11 - \exp\left(\frac{-\pi^2 T_v}{4}\right) \qquad (41)$$

The variation of $U_{av}$ for various values of $T_v$ is given in Figure.

## Numerical Solution for One-Dimensional Consolidation

In this, we will consider the finite-difference solution for one-dimensional consolidation, starting from the basic differential equation of Terzaghi's consolidation theory:

$$\frac{\partial u}{\partial t} = C_v \frac{\partial^2 u}{\partial z^2} \qquad (42)$$

Let $U_R$, $t_R$, $z_R$ by any arbitrary reference excess pore water pressure, time, and distance, respectively. From these we can define the following nondimensional terms:

Nondimensional excess pore water pressure: $\bar{u} = \dfrac{u}{u_R}$ (43)

Nondimensional time: $\bar{t} = \dfrac{t}{t_R}$ (44)

Nondimensional depth: $\bar{z} = \dfrac{z}{z_R}$ (45)

From equations (43), (44), and from the above equation (a),

$$\frac{\partial u}{\partial t} = \frac{u_R}{t_R} \frac{\partial \bar{u}}{\partial \bar{t}} \qquad (46)$$

Similarly, from equations (43), (45), and the right-hand side of equation (42),

$$C_v \frac{\partial^2 u}{\partial z^2} = C_v \frac{u_R}{z_R^2} \frac{\partial^2 \bar{u}}{\partial \bar{z}^2} \tag{47}$$

From equations (46), and (47),

$$\frac{u_R}{t_R} = C_v \frac{\partial \bar{u}}{\partial \bar{t}} \frac{u_R}{z_R^2} \frac{\partial^2 \bar{u}}{\partial \bar{z}^2}$$

Or $\dfrac{1}{t_R} \dfrac{\partial \bar{u}}{\partial \bar{t}} = \dfrac{C_v}{z_R^2} \dfrac{\partial^2 \bar{u}}{\partial \bar{z}^2}$ \hfill (48)

If we adopt the reference time in such a way that $t_R = z_R^2 / C_v$, then equation (48) will be of the form

$$\frac{\partial \bar{u}}{\partial \bar{t}} = \frac{\partial^2 \bar{u}}{\partial \bar{z}^2} \tag{49}$$

The left-hand side of equation (49) can be written as

$$\frac{\partial \bar{u}}{\partial \bar{t}} = \frac{1}{\Delta \bar{t}} (\bar{u}_{0,\bar{t}+\Delta\bar{t}} - \bar{u}_{0,\bar{t}}) \tag{50}$$

Where $\bar{u}_{0,\bar{t}}$ and $\Delta \bar{t}$ are the nondimensional pore water pressure at point O at nondimensional times and $t + \Delta t$.

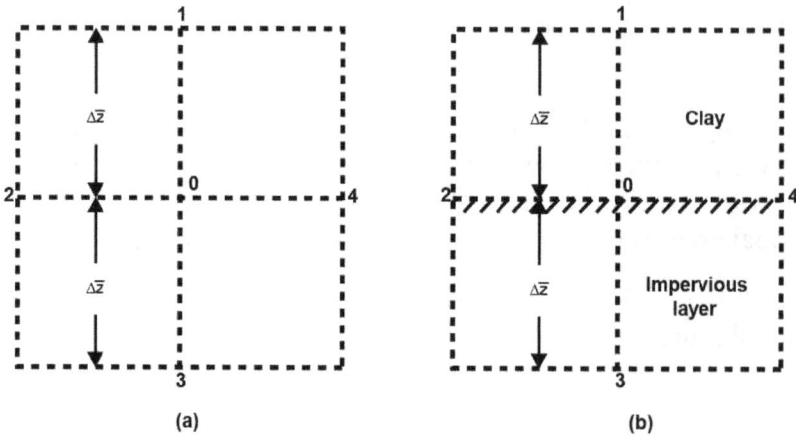

(a)                                                                 (b)

$$\frac{\partial^2 \bar{u}}{\partial \bar{z}^2} = \frac{1}{\Delta \bar{t}} (\bar{u}_{1,\bar{t}} + \bar{u}_{3,\bar{t}} - 2\bar{u}_{0,\bar{t}} \tag{51}$$

Equating the right sides of equations (50) and (51),

$$\frac{1}{\Delta \overline{t}}(\overline{u}_{0,\overline{t}+\Delta \overline{t}} - \overline{u}_{0,\overline{t}}) = \frac{1}{(\Delta \overline{t})^2}(\overline{u}_{1,\overline{t}} + \overline{u}_{3,\overline{t}} - 2\overline{u}_{0,\overline{t}})$$

Or $\overline{u}_{0,\overline{t}+\Delta \overline{t}} = \dfrac{\Delta \overline{t}}{(\Delta \overline{t})^2}(\overline{u}_{1,\overline{t}} + \overline{u}_{3,\overline{t}} - 2\overline{u}_{0,\overline{t}}) + \overline{u}_{0,t}$ \hfill (52)

For equation (52) to coverage, $\Delta \overline{t}$ and $\Delta \overline{z}$ must be chosen such that $\Delta \overline{t}/(\Delta \overline{z})^2$ is less than 0.5.

When solving for pore water pressure at the interface of a clay layer and an impervious layer, equation (52) can be used. However, we need to take point 3 as the mirror image of point 1; thus $\overline{u}_{1,\overline{t}} = \overline{u}_{3,\overline{t}}$. So equation (52) becomes

$$\overline{u}_{0,\overline{t}+\Delta \overline{t}} = \frac{\Delta \overline{t}}{(\Delta \overline{z})^2}(2\overline{u}_{1,\overline{t}} - 2\overline{u}_{0,\overline{t}}) + \overline{u}_{0,\overline{t}} \hfill (53)$$

➢ Consolidation in a layered soil

It is not always possible to develop a closed-form solution for consolidation in layered soils. There are several variables involved, such as different coefficients of permeability, the thickness of layers, and different values of coefficient of consolidation. Figure shows the nature of the degree of consolidation of a two-layered soil.

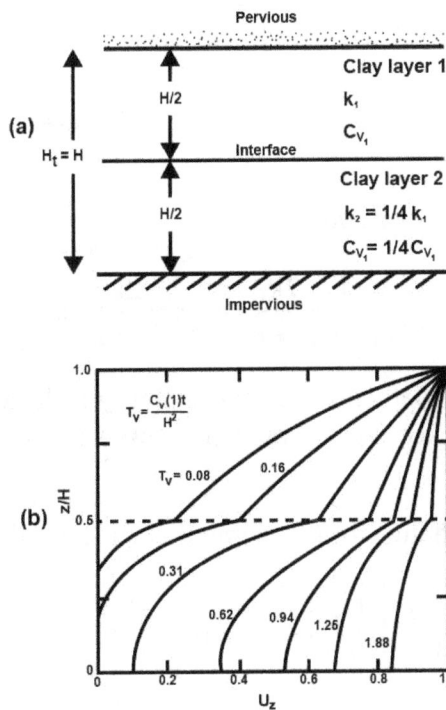

Degree of consolidation in two-layered soil.
(Figure b after U. Luscher, Discussion. Soil Mech. Found. Div., ASCE, vol. 91, no. SM1, 1965)

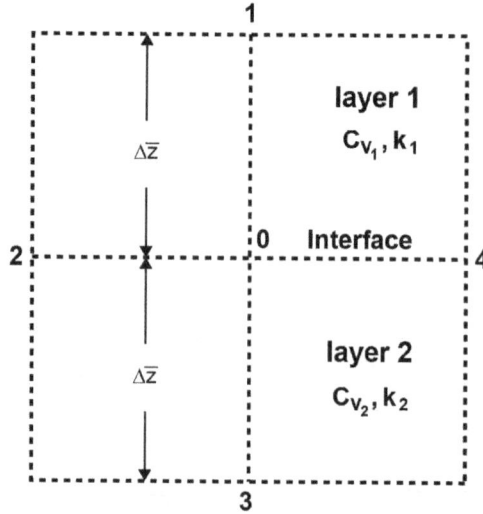

In view of the above, numerical solutions provide a better approach. If we are involved with the calculation of excess pore water pressure at the interface of two different types (i.e., different values of $C_v$) of clayey soils, equation (52) will have to be modified to some extent. Referring to Figure, this can be achieved as follows (Scott, 1963): from equation (a),

$$\frac{k}{C_v}\frac{\partial u}{\partial t} \qquad\qquad = \qquad\qquad K\frac{\partial^2 u}{\partial z^2}$$

$$\uparrow \qquad\qquad\qquad\qquad\qquad\qquad\qquad\qquad \uparrow$$

Change in volume                                   Difference between the rate of
                                                   flow

Based on the derivations of equation

$$k\frac{\partial^2 u}{\partial z^2} = \frac{1}{2}\left[\frac{k_1}{(\Delta z)^2}+\frac{k_2}{(\Delta z)^2}\right]\left(\frac{2k_1}{k_1+k_2}u_{1,t}+\frac{2k_2}{k_1+k_2}u_{3,t}-2u_{0,t}\right) \qquad (54)$$

Where $k_1$ and $k_2$ are the coefficients of permeability in layers 1 and 2, respectively. $u_{0,t}$, $u_{1,t}$ and $u_{3,t}$ are the excess pore water pressures at time t for points 0, 1, and 3, respectively.

Also, the average volume change for the element at the boundary is

$$\frac{k}{C_v}\frac{\partial u}{\partial t} = \frac{1}{2}\left[\frac{k_1}{C_{v1}}+\frac{k_2}{C_{v2}}\right]\frac{1}{\Delta t}(u_{0,t+\Delta t}-u_{0,t}) \qquad (55)$$

Where $u_{0,t}$ and $u_{0,t+\Delta t}$ are the excess pore water pressures at point 0 at times t and $t + \Delta t$, respectively. Equating the right-hand sides of equations (54) and (55), we get

$$\left(\frac{k_1}{C_{v_1}}+\frac{k_2}{C_{v_2}}\right)\frac{1}{\Delta t}(u_{0,t+\Delta t}-u_{0,t})$$

$$=\frac{1}{(\Delta z)^2}(k_1+k_2)\left(\frac{2k_1}{k_1+k_2}u_{1,t}+\frac{2k_2}{k_1+k_2}u_{3,t}-2u_{0,t}\right)$$

Or $u_{0,t+\Delta t}=\dfrac{\Delta t}{(\Delta z)^2}\dfrac{(k_1+k_2)}{k_1/C_{v1}+k_2/C_{v2}}\times\left(\dfrac{2k_1}{k_1+k_2}u_{1,t}+\dfrac{2k_2}{k_1+k_2}u_{3,t}-2u_{0,t}\right)+u_{0,t}$

Or $u_{0,t+\Delta t}=\dfrac{\Delta tCv_1}{(\Delta z)^2}\dfrac{1+k_2/k_1}{1+(k_2/k_1)(C_{v_1}+C_{v_2})}\times\left(\dfrac{2k_1}{k_1+k_2}u_{1,t}+\dfrac{2k_2}{k_1+k_2}u_{3,t}-2u_{0,t}\right)+u_{0,t}$   (56)

Assuming $1/t_R=C_{v_1}/z_R^2$ and combining equations (43) to (45) and (56), we get

$$u_{0,\bar{t}+\Delta\bar{t}}=\frac{1+k_2/k_1}{1+(k_2/k_1)(C_{v_1}+C_{v_2})}\frac{\Delta\bar{t}}{(\Delta\bar{z})^2}\times\left(\frac{2k_1}{k_1+k_2}\bar{u}_{1,\bar{t}}+\frac{2k_2}{k_1+k_2}\bar{u}_{3,\bar{t}}-2\bar{u}_{0,\bar{t}}\right)+u_{0,\bar{t}}$$   (57)

Example 5: A uniform surcharge of $q=150kN/m^2$ is applied at the ground surface of the soil profile shown in Figure. Using the numerical method, determine the distribution of excess pore water pressure for the clay layers after 10 days of load application.

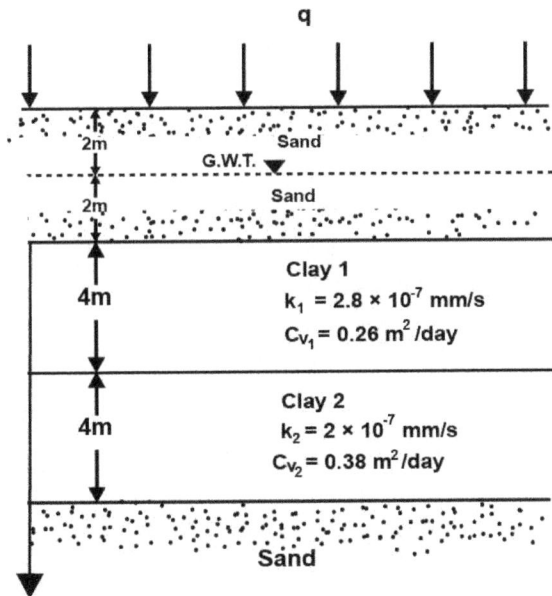

Solution: Since this is a uniform surcharge, the excess pore water pressure immediately after the load application will be $150kN/m^2$ throughout the clay layers. However, due to the drainage conditions, the excess pore water pressures at the top of layer 1 and bottom of layer 2 will immediately become zero. Now, let $z_R = 8m$ and $u_R = 1.5 kN/m^2$. So $\bar{z} = (8m)/(8) = 1$ and $\bar{u} = (150 kN/m^2)/1.5 kN/m^2) = 100$. Figure shows the distribution of $\bar{u}$ at time $t = 0$; note that $\Delta\bar{z} = 2/8 = 0.25$. Now,

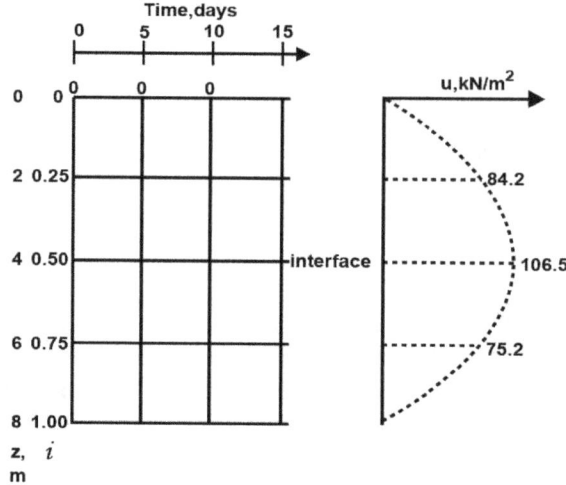

$$t_R = \frac{z_R^2}{C_v} \quad \bar{t} = \frac{t}{t_R} \quad \frac{\Delta t}{\Delta \bar{t}} = \frac{z_R^2}{C_v} \quad \text{or} \quad \Delta\bar{t} = \frac{C_v \Delta t}{z_R^2}$$

Let $\Delta t = 5$ days for both layers. So, for layer 1,

$$\Delta\bar{t}_{(1)} = \frac{C_{v_1}\Delta t}{z_R^2} = \frac{0.26(5)}{8^2} = 0.0203 \quad \frac{\Delta\bar{t}_{(1)}}{(\Delta\bar{z})^2} = \frac{0.0203}{0.25^2} = 0.325 \quad (<0.5)$$

For layer 2,

$$\Delta\bar{t}_{(2)} = \frac{C_{v_2}\Delta t}{z_R^2} = \frac{0.38(5)}{8^2} = 0.0297 \quad \frac{\Delta\bar{t}_{(2)}}{(\Delta\bar{z})^2} = \frac{0.0297}{0.25^2} = 0.475 \quad (<0.5)$$

$$\bar{u} \text{ at } t = 5 \text{ days} : \text{At } \bar{z} = 0$$

$$\bar{u}_{0,\bar{t}+\Delta\bar{t}} = 0$$

At $\bar{z} = 0.25$

$$\bar{u}_{0,\bar{t}+\Delta\bar{t}} = \frac{\Delta\bar{t}_1}{(\Delta\bar{z})^2}(\bar{u}_{1,\bar{t}} + \bar{u}_{3,\bar{t}} - 2\bar{u}_{0,\bar{t}}) + \bar{u}_{0,\bar{t}}$$

At $\bar{z} = 0.5$ [note: this is the boundary of two layers, so we will use equation (57)],

$$\bar{u}_{0,\bar{t}+\Delta\bar{t}} = \frac{1+k_2/k_1}{1+(k_2/k_1)(C_{v_2}+C_{v_1})} \frac{\Delta\bar{t}_{(1)}}{(\Delta\bar{z})^2} \times \left(\frac{2k_1}{k_1+k_2}\bar{u}_{1,\bar{t}} + \frac{2k_2}{k_1+k_2}\bar{u}_{3,\bar{t}} - 2\bar{u}_{0,\bar{t}}\right) + \bar{u}_{0,\bar{t}}$$

$$= \frac{1 + 2/2.8}{1 + (2 \times 0.26)/(2.8/0.38)}(0.325)$$

$$\times \left[ \frac{2 \times 2.8}{2 + 2.8}(100) + \frac{2 \times 2}{2 + 2.8}(100) - 2 \right] + 100$$

Or $\overline{u}_{0,\overline{t}+\Delta\overline{t}} = (1.152)(0.325)(116.67 + 83.33 - 200) + 100 = 100$

At $\overline{z} = 0.75$

$$\overline{u}_{0,\overline{t}+\Delta\overline{t}} = \frac{\Delta\overline{t}_2}{(\Delta\overline{z})^2}(\overline{u}_{1,\overline{t}} + \overline{u}_{3,\overline{t}} - 2\overline{u}_{0,\overline{t}}) + \overline{u}_{0,\overline{t}}$$

$$= 0.475[100 + 0 - 2(100)] + 100 = 52.5$$

At $\overline{z} = 1.0$.

$$\overline{u}_{0,\overline{t}+\Delta\overline{t}} = 0$$

$\overline{u}$ at t = 10 days : At $\overline{z} = 0$

$$\overline{u}_{0,\overline{t}+\Delta\overline{t}} = 0$$

At $\overline{z} = 0.25$

$$\overline{u}_{0,\overline{t}+\Delta\overline{t}} = 0.325[0 + 100 - 2(67.5)] + 67.5 = 56.13$$

$$\overline{z} = 0.5$$

$$\overline{u}_{0,\overline{t}+\Delta\overline{t}} = (1.152(0.325)\left[\frac{2 \times 2.8}{2 + 2.8}(67.5) + \frac{2 \times 2}{2 + 2.8}(52.5) - 2(100)\right] + 100$$

$$= (1.152)(0.325)(78.75 + 43.75 - 200) + 100 = 70.98$$

At $\overline{z} = 0.75$,

$$\overline{u}_{0,\overline{t}+\Delta\overline{t}} = 0.475[100 + 0 - 2(52.5)] + 52.5 = 50.12$$

At $\overline{z} = 1.0$

$$\overline{u}_{0,\overline{t}+\Delta\overline{t}} = 0$$

The variation of the nondimensional excess pore water pressure is shown in Figure. Knowing $\overline{u} = (\overline{u})(u_R) = \overline{u}(1.5)kN/m^2$, we can plot the variation of u with depth.

## Standard One-Dimensional Consolidation Test

The standard one-dimensional consolidation test is usually carried out on saturated specimens about 1 in (25.4 mm) thick and 2.5 in (63.5 mm) in diameter. The soil sample is kept inside a metal ring, with a porous stone at the top and another at the bottom. The load P on the sample is applied through a lever arm, and the compression of the specimen is measured by a micrometer dial gauge. The load is usually doubled every 24 hours. The specimen is kept under water throughout the test.

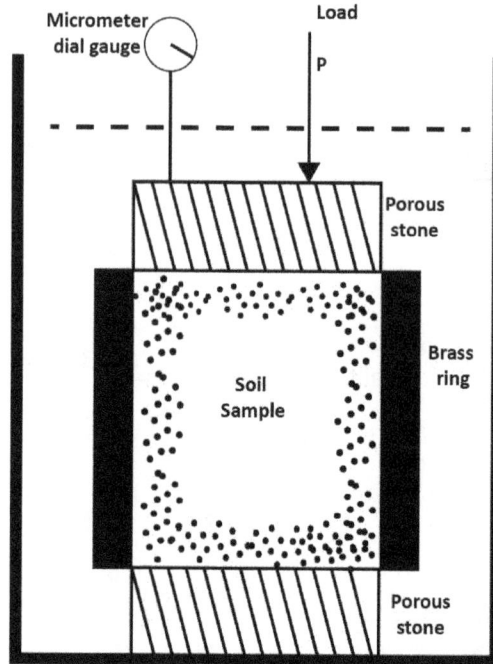

Standard one dimensional consolidation apparatus

For each load increment, the sample deformation and the corresponding time t is plotted on semilogarithmic graph paper. Figure shows a typical deformation vs. log t graph. The graph consists of three distinct parts:

1. Upper curved portion (stage I). This is mainly the result of precompression of the specimen.

2. A straight-line portion (stage II). This is referred to as primary consolidation. At the end of the primary consolidation, the excess pore water pressure generated by the incremental loading is dissipated to a large extent.

3. A lower straight-line portion (stage III). This is called secondary consolidation. During this stage, the specimen undergoes small deformation with time. in fact, there must be immeasurably small excess pore water pressure in the specimen during secondary consolidation.

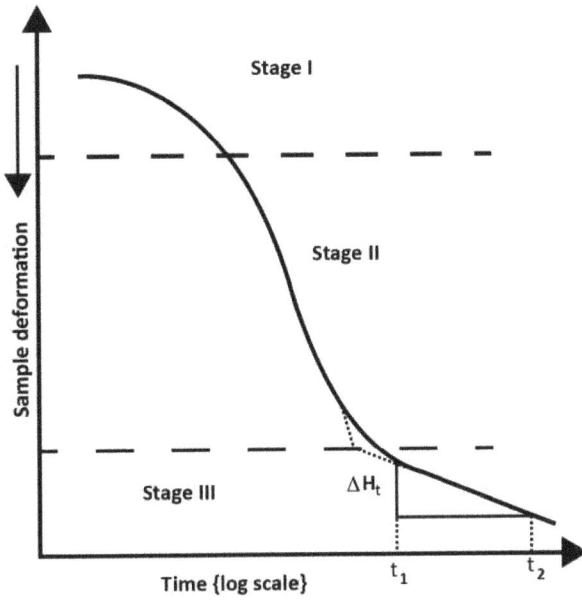

Typical sample deformation vs. log-of-time plot for a given load increment

Note that at the end of the test for each incremental loading the stress on the specimen is the effective stress, $\sigma'$. Once the specific gravity of the soil solids, the initial specimen dimensions, and the specimen deformation at the end of each load has been determined, the corresponding void ratio can be calculated. A typical void ratio vs. effective pressure relationship plotted on semilogarithmic graph paper is shown in Figure.

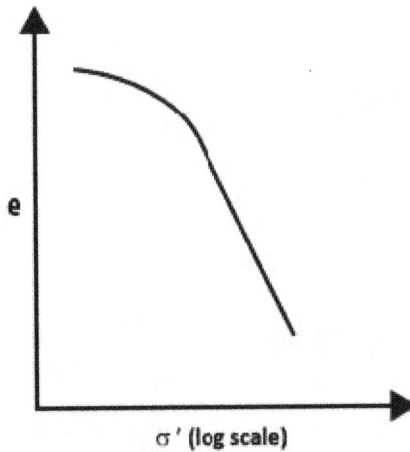

Typical e vs. log plot

## Calculation of One-dimensional Consolidation Settlement

The basic principle of one-dimensional consolidation settlement calculation is demonstrated in Figure. If a clay layer of total thickness $H_t$ is subjected to an increase of average effective overburden pressure from $\sigma_0'$ to $\sigma_1'$, it will undergo a consolidation settlement of $\Delta H_t$. Hence the strain can be given by

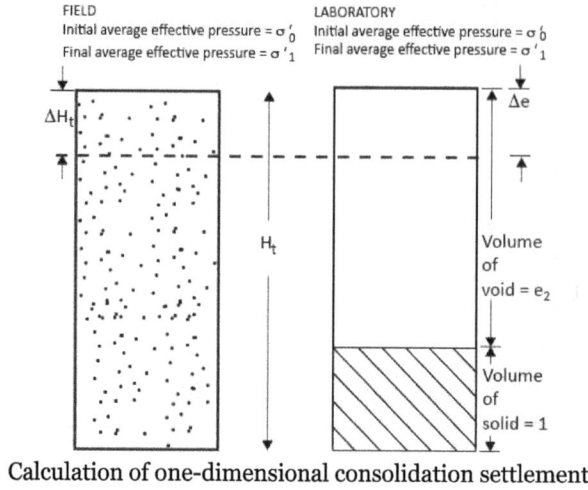

FIELD
Initial average effective pressure = $\sigma'_0$
Final average effective pressure = $\sigma'_1$

LABORATORY
Initial average effective pressure = $\sigma'_0$
Final average effective pressure = $\sigma'_1$

Calculation of one-dimensional consolidation settlement

$$\underline{\quad\quad\quad}$$ (58)

Where e is strain. Again, if an undisturbed laboratory specimen is subjected to the same effective stress increase, the void ratio will decrease by $\Delta e$. Thus, the strain is equal to

$$\epsilon = \frac{\Delta e}{1 + e_0} \qquad (59)$$

Where e is the void ratio at an effective stress of $\sigma'_0$.

Thus, from equations (58) and (59),

$$\Delta H_t = \frac{\Delta e H_t}{1 + e_0} \qquad (60)$$

For a normally consolidated clay in the field,

$$\Delta e = C_c \log \frac{\sigma'_1}{\sigma'_0} = C_c \log \frac{\sigma'_0 + \Delta \sigma}{\sigma'_0} \qquad (61)$$

For an overconsoidated clay, (1) if $\sigma'_1 < \sigma'_c$ (i.e., overconsolidated pressure )

$$\Delta e = C_r \log \frac{\sigma'_1}{\sigma'_0} = C_r \log \frac{\sigma'_0 + \Delta \sigma}{\sigma'_0}$$

And (2) if $\sigma'_0 < \sigma'_c < \sigma'_1$

$$\Delta e = \Delta e_1 + \Delta e_2 = C_r \log \frac{\sigma'_c}{\sigma'_0} + C_r \log \frac{\sigma'_0 + \Delta \sigma}{\sigma'_c}$$

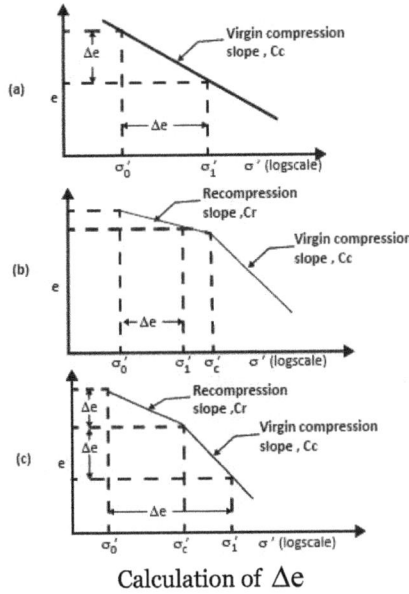

Calculation of $\Delta e$

## Constant Rate-of-Strain Consolidation Tests

The standard one-dimensional consolidation test procedure is time-consuming. Recently, at least two other one-dimensional consolidation test procedure have been developed which are much faster but yet give reasonably good results. The methods are (1) the constant rate-of-strain consolidation test and (2) the constant-gradient consolidation test.

The constant rate-of-strain method was developed by smith and Wahls (1969). A soil specimen is taken in a fixed-ring consolidometer and saturated. For conducting the test, drainage is permitted at the top of the sample, but not at the bottom. A continuously increasing load is applied to the top of the specimen so as to produce a constant rate of compressive strain, and the excess pore water pressure $u_b$ (generated by the continuously increasing stress $\sigma$ at the top) is measured. Figure shows a schematic diagram of the laboratory test setup.

Theory. The mathematical derivations developed by Smith and Wahls for obtaining the void ratio-effective pressure relationship and the corresponding coefficient of consolidation are given below.

The basic equation for continuity of flow through a soil element is given as

$$\frac{k}{\gamma_w}\frac{\partial^2 u}{\partial z^2} = \frac{1+\partial e}{1+e\,\partial t}$$

The coefficient of permeability at a given time is a function of the average void ratio $\bar{e}$ in the specimen. The The average void ratio is, however, continuously changing due to

the constant rate of strain. Thus,

$$k = k(\overline{e}) = f(t) \qquad (62)$$

The average void ratio is given by

$$\overline{e} = \frac{1}{H} \int_0^H e\,dz$$

Where $H(= H_t)$ is the sample thickness.

In the constant rate-of-stain type of test, the rate of change of volume is constant, or

$$\frac{dv}{dt} = -RA \qquad (63)$$

Where

V = volume of specimen

A = area of cross section of specimen

R = constant rate of deformation of upper surface

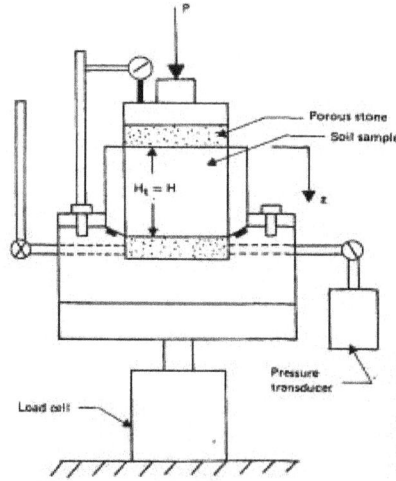

Schematic diagram for laboratory test setup for
controlled rate-of-strain type of test (After Smith and Wahls, 1969)

The rate of change of average void ratio $\overline{e}$ can be given by

$$\frac{d\overline{e}}{dt} = \frac{1}{V_s}\frac{dV}{dt} = -\frac{1}{V_s}RA = -r \qquad (64)$$

Where r is a constant

Based on the definition of $\overline{e}$ and equation (62), we can write

$$e_{(z,f)} = g(z)t + e_0 \qquad (65)$$

Where

$e_{(z,f)}$ = void ratio at depth $z$ and time $t$

$e_0$ = initial void ratio at beginning of test

$g(z)$ = a function of depth only

The function $g(z)$ is difficult to determine. We will assume it to be a linear function of the form

$$-r\left[1 - \frac{b}{r}\left(\frac{z - 0.5H}{H}\right)\right]$$

Where $b$ is a constant, substitution of this into equation (65) gives

$$e_{(z,f)} = e_0 - rt\left[1 - \frac{b}{r}\left(\frac{z - 0.5H}{H}\right)\right] \qquad (66)$$

Let us consider the possible range of variation of $b/r$ as given in equation (66):

1. $b/r = 0$

$$e_{(z,f)} = e_0 - rt \qquad (67)$$

    This indicates that the void is constant with depth and changes with time only. In reality, this is not the case.

2. If $b/r = 2$ the void ratio at the base of the sample, i.e., at $z = H$, becomes

$$e_{(z,f)} = e_0 \qquad (68)$$

    This means that the void ratio at the base does not change with time at all, which is not realistic.

So the value of $b/r$ is somewhere between 0 and 2 and may be taken as about 1.

Assuming $b/r \neq 0$ and using the definition of void ratio as given by equation (66), we can integrate earlier equation to obtain an equation for the excess pore water pressure. The boundary conditions are: at $z = 0$, $u = 0$ (at any time); and at $z = H$, $\partial u / \partial z = 0$ (at any time). Thus, o

$$u = \frac{\gamma_w r}{k}\left\{zH\left[\frac{1 + e_0 - bt}{rt(bt)}\right] + \frac{z^2}{2rt} - \left[\frac{H(1 + e_0)}{rt(bt)}\right] \times \left[H\frac{(1 + e)}{bt}\ln(1 + e) - z\ln(1 + e_B) - \frac{H(1 + e_T)}{bt}\ln(1 + e_T)\right]\right\} \qquad (69)$$

Where $e_B = e_0 - rt\left(1 - \frac{1}{2}\frac{b}{r}\right) \qquad (70)$

$$e_T = e_0 - rt\left(1 + \frac{1}{2}\frac{b}{r}\right) \tag{71}$$

Equation (69) is very complicated. Without loosing a great deal of accuracy, it is possible to obtain a simpler form of expression for u by assuming that the term $1+\bar{e}$ (note that this is not a function of z). so, from equations (65) and (66),

$$\frac{\partial^2 u}{\partial z^2} = \left[\frac{\gamma_w}{k(1+\bar{e})}\right]\frac{\partial}{\partial t}\left\{e_0 - rt\left[1 - \frac{b}{r}\left(\frac{z-0.5H}{H}\right)\right]\right\} \tag{72}$$

Using the boundary condition u = 0 at z = 0 and $\partial u/\partial t$ = 0 at z = H, equation (98) can be integrated to yield

$$u = \left[\frac{\gamma_w r}{k(1+\bar{e})}\right]\left[\left(Hz - \frac{z^2}{2}\right) - \frac{b}{r}\left(\frac{z^2}{4} - \frac{z^3}{6H}\right)\right] \tag{73}$$

The pore pressure at the base of specimen can be obtained by substituting z = H in equation(73)

$$u_{z=H} = \frac{\gamma_w r H^2}{k(1+\bar{e})}\left(\frac{1}{2} - \frac{1}{12}\frac{b}{r}\right) \tag{74}$$

The average effective stress corresponding to a given value of $u_{z=H}$ can be obtained by writing

$$\sigma'_{av} = \sigma - \frac{U_{av}}{u_{z=H}}u_{z=H} \tag{75}$$

Where

$\sigma'_{av}$ = average effective stress on specimen at any time

$\sigma$ = total stress on sample

$U_{av}$ =corresponding average pore water pressure

$$\frac{U_{av}}{u_{z=H}} = \frac{\frac{1}{H}\int_0^H u\,dz}{u_{z=H}} \tag{76}$$

Substitution of equations (73) and (1000 into equation (76) and further simplification gives

$$\frac{U_{av}}{u_{z=H}} = \frac{\frac{1}{3} - \frac{1}{24}(b/r)}{\frac{1}{2} - \frac{1}{12}(b/r)} \tag{77}$$

Note that for $(b/r) = 0$, $U_{av}/u_{z=H} = 0.667$; and for $(b/r) = 1$, $U_{av}/u_{z=H} = 0.700$. Hence, for $0 \leq b/r \leq 1$, the values of $U_{av}/u_{z=H}$ does not change significantly. So, form equations (75) and (77),

$$\sigma'_{av} = \sigma - \left[ \dfrac{\dfrac{1}{3} - \dfrac{1}{24}(b/r)}{\dfrac{1}{2} - \dfrac{1}{12}(b/r)} \right] u_{z=H} \tag{78}$$

➤ Coefficient of consolidation

The coefficient of consolidation was defined previously as

$$C_v = \frac{k(1+e)}{a_v \gamma_w}$$

We can assume $1 + e \approx 1 + \overline{e}$, and from equation (100)

$$k = \frac{\gamma w^r H^2}{(1+\overline{e})u_{z=H}} \left( \frac{1}{2} - \frac{1}{12} \frac{b}{r} \right) \tag{79}$$

Substitution of these into the expression for $C_v$ gives

$$C_v = \frac{r H^2}{a_v u_{z=H}} \left( \frac{1}{2} - \frac{1}{12} \frac{b}{r} \right) \tag{80}$$

➤ Interpretation of experimental results

The following information can be obtained from a constant rate-of-strain consolidation test:

1. Initial height of sample, $H_i$.

2. A.

3. $V_s$.

4. Stain rate R.

5. A continuous record of $u_{z=H}$.

6. A corresponding record of $\sigma$ (total stress applied at the top of the specimen).

The plot of e vs. $\sigma'_{av}$ can be obtained in the following manner:

1. Calculate $r = RA / V_s$.

2. Assume $b / r \approx 1$.

3. For a given value of $u_{z=H}$, the value of $\sigma$ is known (at time t from the start of the test), and so $\sigma'_{av}$ can be calculated from equation (78).

4. Calculate $\Delta H = Rt$ and then the change in void ratio that has taken place curing time t.

$$\Delta e = \frac{\Delta H}{H_i}(1 + e_0)$$

Where $H_i$ is the initial height of the sample.

5. The corresponding void ratio (at time t) is $e = e_0 - \Delta e$.

6. After obtaining a number of points of $\sigma'_{av}$ and the corresponding e, plot the graph of e vs. $\log \sigma'_{av}$.

7. For a given value of $\sigma'_{av}$ and e, the coefficient of consolidation $C_v$ can be calculated by using equation (80). (Note that H in equation (80) is equal to $H_i - \Delta H$

.

## One-Dimensional Consolidation with Visoelastic Models

The rheological model for soil chosen by Barden consists of a linear spring and nonlinear dashpot as shown in Figure. The equation of continuity for one-dimensional consolidation is given as

$$\frac{k(1+e)}{\gamma_w} \frac{\partial^2 u}{\partial z^2} = \frac{\partial e}{\partial t}$$

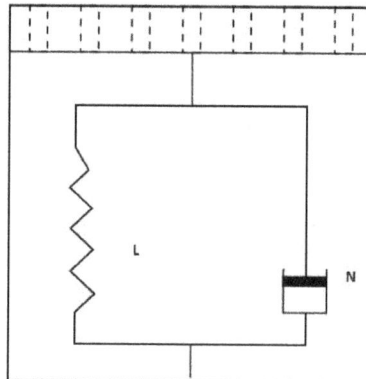

Rheological model for soil. L: Linear spring; N: Nonlinear dashpot

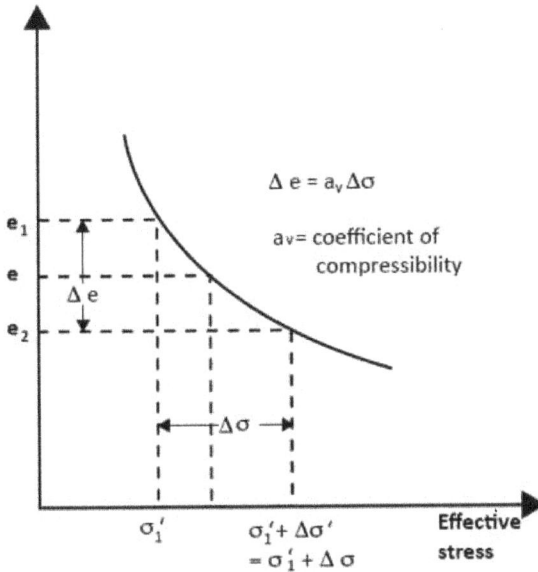

Nature of variation of void ratio with effective stress

From Figure,

$$\frac{e_1 - e_2}{a_v} = \frac{e_1 - e}{a_v} + u + \tau \tag{81}$$

Where $\dfrac{e_1 - e_2}{a_v} = \Delta\sigma'$ = total effective stress increase the soil will be subjected to at end of consolidation

$\dfrac{e_1 - e}{a_v}$ = Effective stress increase in the soil at some stage of consolidation

(i.e., the stress carried by the soil grain bond, represented by the spring in Figure)

u = excess pore water pressure

 = strain carried by film bond (represented by the dashpot in Figure)

The strain $\tau$ can be given by a power-law relation:

$$r = b\left(\frac{\partial e}{\partial t}\right)^{1/n}$$

Where $n > 1$, and b is assumed to be a constant over the pressure range $\Delta\sigma$. Substitution of the preceding power-law relation for $\tau$ in equation (81) and simplification gives

$$e - e_2 = a_v\left[u + b\left(\left(\frac{\partial e}{\partial t}\right)^{1/n}\right)\right] \tag{82}$$

Now let $e - e_2 = e'$ So,

$$\frac{\partial e'}{\partial t} = \frac{\partial e}{\partial t} \tag{83}$$

$$\bar{z} = \frac{z}{H} \tag{84}$$

Where H is the length of maximum drainage path, and

$$\bar{u} = \frac{u}{\Delta\sigma'} \tag{85}$$

The degree of consolidation is

$$U_Z = \frac{e_1 - e}{e_1 - e_2} \tag{86}$$

And $\lambda = 1 - U_Z = \dfrac{e - e_2}{e_1 - e_2} = \dfrac{e'}{a_v \Delta\sigma'}$ $\qquad$ (87)

Elimination of u from equations (9) and (123) yields

$$\frac{k(1+e)}{\gamma_w} \frac{\partial^2}{\partial z^2} \left[ \frac{e'}{a_v} - b \left( \frac{\partial e'}{\partial t} \right)^{1/n} \right] = \frac{\partial e'}{\partial t} \tag{88}$$

Combining equations (125), (128), and (129) we obtain

$$\frac{\partial^2}{\partial z^2} \left\{ \lambda - \left[ a_v b^n (\Delta\sigma')^{1/n} \frac{\partial\lambda}{\partial t} \right]^{1/n} \right\} = \frac{a_v H^2 \gamma_w}{k(1+e)} \frac{\partial\lambda}{\partial t} = \frac{m_v H^2 \gamma_w}{k} \frac{\partial\lambda}{\partial t} = \frac{H^2}{C_v} \frac{\partial\lambda}{\partial t} \tag{89}$$

Where $m_v$ is the volume coefficient of compressibility and $C_v$ is the coefficient of con-solidation.

The right-hand side of equation (89) can be written in the form

$$\frac{\partial\lambda}{\partial T_V} = \frac{H^2}{C_v} \frac{\partial\lambda}{\partial t} \tag{90}$$

Where $T_V$ is the nondimensional time factor and is equal to $C_v t / H^2$.

Similarly defining

$$T_S = \frac{t(\Delta\sigma')^{n-1}}{a_v b^n} \tag{91}$$

We can write

$$\left[a_v b^n (\Delta\sigma')^{1-n} \frac{\partial\lambda}{\partial t}\right]^{1/n} = \left(\frac{\partial\lambda}{\partial T_S}\right)^{1/n} \tag{92}$$

$T_S$ in equations (91), and (92) is defined as structural viscosity.

It is useful now to define a nondimensional ratio R as

$$R = \frac{T_v}{T_S} = \frac{C_v a_v}{H^2} \frac{b^n}{(\Delta\sigma')^{n-1}} \tag{93}$$

From equations (89), (90), and (92),

$$\frac{\partial^2}{\partial z^2}\left[\lambda - \left(\frac{\partial\lambda}{\partial T_S}\right)^{1/n}\right] = \frac{\partial\lambda}{\partial T_v} \tag{94}$$

Note that equation (92) in nonlinear. For that reason, Barden suggested solving the two simultaneous

equation obtained from the basic earlier equation.

$$\frac{\partial^2\bar{u}}{\partial z^2} = \frac{\partial\lambda}{\partial T_v} \tag{95}$$

And $-\dfrac{1}{R}(\lambda - \bar{u})^n = \dfrac{\partial\lambda}{\partial T_v}$ $\tag{96}$

# Secondary Consolidation

Several investigations have been carried out for qualitative and quantitative evaluation of secondary consolidation. The magnitude of secondary consolidation is often defined by

$$c_\alpha = \frac{\Delta H_t / H_t}{\log t_2 - \log t_1} \tag{97}$$

Where $c_\alpha$ is the coefficient of secondary consolidation.

In order to study the effect of remolding and preloading on secondary compression, Mesri (1973) conducted a series of one-dimensional consolidation tests on organic Paulding clay. Figure shows the results in the form of a plot of $\Delta e / (\Delta \log t)$ vs. with load increment ratios of 1 and with only sufficient time allowed for excess pore water pressure dissipation. Under the final pressure, secondary compression was observed for a period of 6 months. The following conclusions can be drawn from the results of these tests:

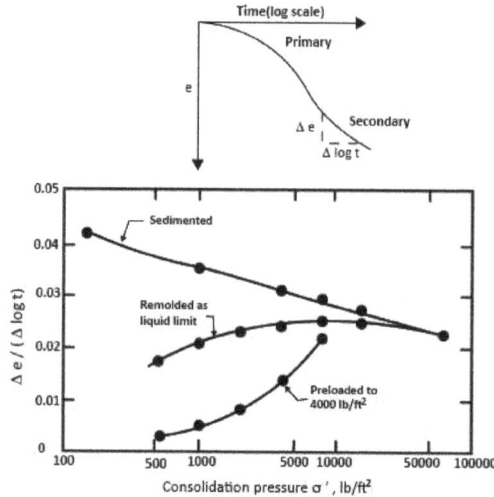

Coefficient of secondary compression for organic Paulding clay.
(Note:1 $lb/ft^2$ = 47.9$N/m^3$ )(Redrawn after G. Mesri, Coefficient of
Secondary Compression, J. Soil Mech. Found. Div., ASCE, vol. 99, no.SM1, 1973)

1.   For sedimented (undisturbed) soils, $\Delta e/(\Delta \log t)$ decreases with the increase of the final consolidation pressure.

2.   Remolding of clays creates a more dispersed fabric. This results in a decrease of the coefficient of secondary consolidation at lower consolidation pressures as compared to that for undisturbed samples. However, it increases with consolidation pressure to a maximum value and then decreases finally merging with the values for normally consolidated undisturbed samples.

3.   Precompressed clays show a smaller value of coefficient of secondary consolidation. The degree of reduction appears to be a function of the degree of precompression.

# Constant-Gradient Consolidating Test

The constant-gradient consolidation test was developed by Lowe et al. (1969). In this procedure, a saturated soil sample is taken in a consolidation ring. As in the case of the constant rate-of-strain type of test, drainage is allowed at the top of the sample and pore water pressure is measured at the bottom. A load P is applied on the sample which increases the excess pore water pressure in the specimen by an amount $\Delta u$ (Figure a).

After a small lapse of time $t_1$, the excess pore water pressure at the top of the sample will be equal to zero (since drainage is permitted). However, at the bottom of the sample the excess pore water pressure will still be approximately $\Delta u$ (Figure b). From this point on, the load P is increased slowly in such a way that the difference between the pore water pressure at the top and bottom of the specimen remains constant, i.e., the difference is maintained at a constant $\Delta u$ (Figure c and d). When the desired value of P is reached, say at time $t_3$ the loading is stopped and the excess pore water pressure is allowed to dissipate. The elapsed time $t_4$ at which the pore water pressure at the bottom of the specimen reaches a value of 0.1 $\Delta u$ is recorded. During the entire test, the compression $\Delta H$ that the specimen undergoes is recorded.

Theory: From the basic equation we have

$$\frac{k}{\gamma_w} \frac{\partial^2 u}{\partial z^2} = -\frac{a_v}{1+e} \frac{\partial \sigma'}{\partial t} \qquad (98)$$

Or $\dfrac{\partial \sigma'}{\partial t} = -\dfrac{k}{\gamma_w m_v} \dfrac{\partial^2 u}{\partial z^2} = -C_v \dfrac{\partial^2 u}{\partial z^2}$ $\qquad (99)$

Since $\sigma' = \sigma - u$,

$$\frac{\partial \sigma'}{\partial t} = \frac{\partial \sigma}{\partial t} - \frac{\partial u}{\partial t} \qquad (100)$$

For the controlled-gradient tests (i.e., during the time $t_1$ to $t_3$ in Figure), $\partial u / \partial t = 0$. So,

$$\frac{\partial \sigma'}{\partial t} = \frac{\partial \sigma}{\partial t} \qquad (101)$$

Combining equations (108) and (110),

$$\frac{\partial \sigma}{\partial t} = -C_v \frac{\partial^2 u}{\partial z^2} \qquad (102)$$

Note that the left-hand side of equation (111) is independent of the variable z and the right-hand side is independent of the variable t. so both sides should be equal to a constant, say $A_1$. Thus,

$$\frac{\partial \sigma}{\partial t} = A_1 \qquad (103)$$

And $\qquad \dfrac{\partial^2 u}{\partial z^2} = -\dfrac{A_1}{C_v}$ $\qquad (104)$

Integration of equation (87) yields

$$\frac{\partial u}{\partial z} = -\frac{A_1}{C_v}z + A_2 \tag{105}$$

And $u = -\dfrac{A_1}{C_v}\dfrac{z^2}{2} + A_2 z + A_3$ (106)

The boundary conditions are as follows:

1.  At $z = 0$, $\partial u / \partial z = 0$.

2.  At $z = H$, $u = 0$.

3.  At $z = 0$, $u = \Delta u$.

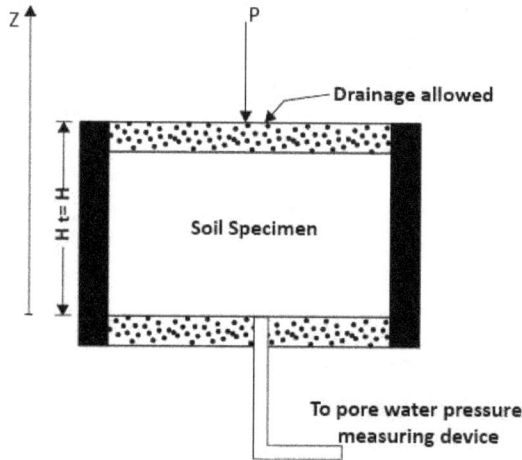

Schematic diagram for constant-gradient consolidation test

From the first boundary condition and equation (105), we find that $A_2 = 0$. So,

$$u = -\frac{A_1}{C_v}\frac{z^2}{2} + A_3 \tag{107}$$

From the secondary boundary condition and equation (107),

$$A_3 = \frac{A_1 H^2}{2C_v} \tag{108}$$

Or $u = \dfrac{A_1}{C_v}\dfrac{z^2}{2} + \dfrac{A_1}{C_v}\dfrac{H^2}{2}$ (109)

For the third boundary condition and equation (109),

$$\Delta u = \frac{A_1}{C_v}\frac{H^2}{2}$$

Or $A_1 = \dfrac{2C_v \Delta u}{H^2}$ (110)

Substitution of this value of $A_1$ into equation (109) yields

$$u = \Delta u \left( 1 - \dfrac{z^2}{H^2} \right)$$ (111)

Equation (111) shows a parabolic pattern of excess pore water pressure distribution, which remains constant during the controlled-gradient test (time $t_1$ to $t_3$ in Figure).

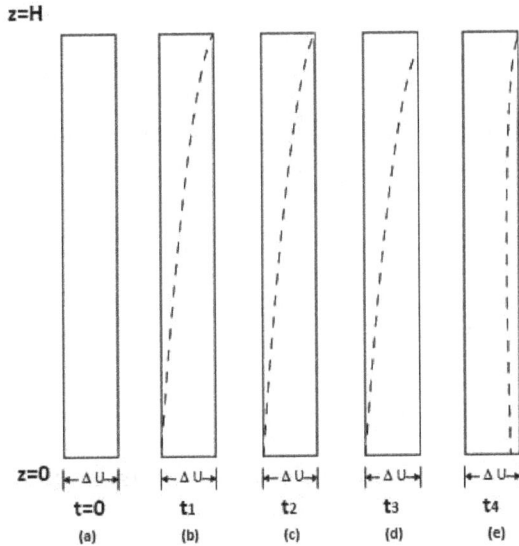

Stages in controlled-gradient test. (Lowe et al. 1965)

Combining equations (112) and (119), we obtain

$$\dfrac{\partial \sigma}{\partial t} = A_1 = \dfrac{2C_v \Delta u}{H^2}$$

Or $C_v = \dfrac{\partial \sigma}{\partial t} \dfrac{H^2}{2\Delta u}$ (112)

> Interpretation of experimental results

The following information will be available from the constant-gradient test:

1. Initial height of the sample, $H_i$, and height, H at any time during the test.

2. The rate of application of the load P and thus the rate of application of stress $\partial \sigma / \partial t$ on the sample.

3. The differential pore pressure $\nabla u$.

4. Time $t_1$.

5. Time $t_3$.

6. Time $t_4$.

The plot of e vs. $\sigma'_{av}$ can be obtained in the following manner:

1. Calculate the initial void ratio $e_0$.

2. Calculate the change in void ratio at any other time t during the test as

$$\Delta e = \frac{\Delta H}{H_i}(1+e_0)$$

(113)

Where $\Delta H$ is the total change in height from the beginning of test. So, the average void ratio at time t is $e = e_0 - \Delta e$.

3. Calculate the average effective stress at time t using the known total stress $\sigma$ applied on the sample at that time:

$$\sigma'_{av} = \sigma - U_{av}$$

(114)

Where $U_{av}$ is the average excess pore water pressure in the sample

Calculation of the coefficient of consolidation is as follows:

1. At time $t_1$,

$$C_v = \frac{0.008H^2}{t_1}$$

(115)

2. At time $t_1 < 1 < t_3$,

$$C_v = \frac{\Delta\sigma}{\Delta t}\frac{H^2}{2\Delta u}$$

(116)

Note that $\Delta\sigma/\Delta t, H,$ and $\Delta u$ are all oknown from the tests.

3. Between time $t_3$ and $t_4$,

$$C_v = \frac{(1.1-0.08)H^2}{t_3 - t_4} = \frac{1.02H^2}{t_3 - t_4}$$

(117)

# Permissions

# Index

www.ingramcontent.com/pod-product-compliance
Lightning Source LLC
Chambersburg PA
CBHW062004190326

41458CB00009B/2963

# Recent Research in Greenhouse Gases